Urban Ecology and Intervention in the 21st Century Americas

This book takes a hemispheric approach to contemporary urban intervention, examining urban ecologies, communication technologies, and cultural practices in the twenty-first century. It argues that governmental and social regimes of control and forms of political resistance converge in speculation on disaster and that this convergence has formed a vision of urban environments in the Americas in which forms of play and imaginations of catastrophe intersect in the vertical field.

Schifani explores a diverse range of resistant urban interventions, imagining the city as on the verge of or enmeshed in catastrophe. She also presents a model of ecocriticism that addresses aesthetic practices and forms of play in the urban environment. Tracing the historical roots of such tactics as well as mapping their hopes for the future will help the reader to locate the impacts of climate change not only on the physical space of the city, but also on the epistemological and aesthetic strategies that cities can help to engender.

This book will be of great interest to students and scholars of Urban Studies, Media Studies, American Studies, Global Studies, and the broad and interdisciplinary field of Environmental Humanities.

Allison M. Schifani is Assistant Professor of Digital Humanities, Department of Modern Languages and Literatures, University of Miami, Florida, USA.

Routledge Environmental Humanities

Series editors: Scott Slovic (University of Idaho, USA), Joni Adamson (Arizona State University, USA) and Yuki Masami (Aoyama Gakuin University, Japan)

The *Routledge Environmental Humanities* series is an original and inspiring venture recognising that today's world agricultural and water crises, ocean pollution and resource depletion, global warming from greenhouse gases, urban sprawl, overpopulation, food insecurity and environmental justice are all *crises of culture.*

The reality of understanding and finding adaptive solutions to our present and future environmental challenges has shifted the epicenter of environmental studies away from an exclusively scientific and technological framework to one that depends on the human-focused disciplines and ideas of the humanities and allied social sciences.

We thus welcome book proposals from all humanities and social sciences disciplines for an inclusive and interdisciplinary series. We favour manuscripts aimed at an international readership and written in a lively and accessible style. The readership comprises scholars and students from the humanities and social sciences and thoughtful readers concerned about the human dimensions of environmental change.

Urban Ecology and Intervention in the 21st Century Americas

Verticality, Catastrophe, and the Mediated City

Allison M. Schifani

Routledge
Taylor & Francis Group
LONDON AND NEW YORK

earthscan
from Routledge

First published 2021
by Routledge
2 Park Square, Milton Park, Abingdon, Oxon OX14 4RN

and by Routledge
52 Vanderbilt Avenue, New York, NY 10017

Routledge is an imprint of the Taylor & Francis Group, an informa business

British Library Cataloguing-in-Publication Data
A catalogue record for this book is available from the British Library

Library of Congress Cataloging-in-Publication Data
Names: Schifani, Allison M., author.
Title: Urban ecology and intervention in the 21st century Americas : verticality, catastrophe, and the mediated city / Allison M. Schifani.
Description: Milton Park, Abingdon, Oxon ; New York, NY : Routledge, 2021. | Series: Routledge environmental humanities | Includes bibliographical references and index. |
Identifiers: LCCN 2020032724 (print) | LCCN 2020032725 (ebook) | ISBN 9780367519360 (paperback) | ISBN 9780367519377 (hardback) | ISBN 9781003055747 (ebook)
Subjects: LCSH: Urban ecology (Sociology)--America--History--21st century. | Urban policy--America--History--21st century.
Classification: LCC HT243.A45 S35 2021 (print) | LCC HT243.A45 (ebook) | DDC 307.76097009/05--dc23
LC record available at https://lccn.loc.gov/2020032724
LC ebook record available at https://lccn.loc.gov/2020032725

ISBN: 978-0-367-51937-7 (hbk)
ISBN: 978-0-367-51936-0 (pbk)
ISBN: 978-1-003-05574-7 (ebk)

Typeset in Bembo
by SPi Global, India

A version of Chapter 2 first appeared as "Alternative Sprawls, Junkcities: Buenos Aires Libre and Horizontal Urban Epistemologies" in *The Journal of Urban Cultural Studies* 1.3, September, 2014. The JUCS generously granted permission to use it here.

For Taos and Milne: Let's build the city together

Contents

Figures

Acknowledgments

This book was made in concert, in collaboration, and in camaraderie with many. My colleagues at the University of Miami were supporters throughout, but particular thanks must go to both Lillian Manzor and George Yudice who helped shepherd me into the department and through these past several years in Miami. They were champions of my work and joyous comrades too.

My research benefited from the generous support of the University of Miami's Center for the Humanities, which granted me time to think and write and led me to colleagues across the College of Arts and Sciences who guided my work and offered constructive feedback.

Lisa Swanstrom gave me thorough, attentive edits and offered provocations and connections that were absolutely essential. Rita Raley guided me as a student and pushed me as a scholar. This book would not be possible without her pedagogical and personal generosity. Allison Carruth, Maurizia Boscagli, and Lisa Parks were inspiring teachers and mentors, and they read some of the work herein when it was barely legible.

My dear friends Jessica Rosenberg and Lindsay Thomas made Miami feel like home, pushed me to clarify my thinking, and fed me delicious dinners. Danai Verde and Tyler Miller were always spry and willing interlocutors and made pandemic writing tolerable with their friendship and their gardening skills. Anna Vernon always knew I could do it.

Bill Deal drank with me and put things in perspective.

David Lyttle argued with me, edited my work, and supported me. Our late night conversations on porches, stoops, and small couches made me, and this book, better.

Finally, I owe a debt of gratitude to my family. My mother's curiosity and close reading helped me here, as always. My sister's cheerleading steeled me. My father's laughter brightened even the darkest of days. Thank you.

1 The weather of catastrophe and the vertical field

Every autumn Los Angeles and the bulk of Southern California are pummeled by the caustically dry Santa Ana winds. Carrying dust and fire-starting heat, these annual winds have their own mythology in the city, where this book first began to germinate many years ago. Sometimes called "the devil winds," the Santa Ana have been said to drive people mad, to induce rage, to spark violence among the city of angels' overheated, world-weary inhabitants. Joan Didion saw them as directly linked to the city's ethos:

> Los Angeles weather is the weather of catastrophe, of apocalypse, and, just as the reliably long and bitter winters of New England determine the way life is lived there, so the violence and unpredictability of the Santa Ana affect the entire quality of life in Los Angeles, accentuate its impermanence, its unreliability. The wind shows us how close to the edge we are.[1]

This is not a book about the Santa Ana winds, but it is a book about catastrophe and the city, as it is imagined by artists, scholars, activists, and other urban wanderers across three cities of the Americas: Buenos Aires, Miami, and Los Angeles. It is a book about the catastrophes already unfolding in these cities, and those so many of us imagine, or perhaps know, are coming.

The impermanence and unreliability Didion understood to be a quality of life specific to Los Angeles has become (if it was not already) the quality of a trans-American urban ethos. Cities are consistently understood to be in dangerously precarious positions in terms not only of shifting and severe weather in the age of anthropogenic climate change, but also in political, social, spatial, and economic terms. When we see the next economic crash on the horizon, we know what it will look like in the city: packed slums, abandoned construction sites, blight in once thriving neighborhoods. We have already seen flood waters rush through urban areas to devastating effect, and hurricanes strengthened by warming ocean waters inundate cities in the Caribbean and across the Eastern seaboard of the United States. Catastrophe is not just a future possibility for cities like New Orleans or San Juan. Nor for the other cities that are the triangulated sites explored over the course of this book. We've already lived through it, are living through it, or at the very least, have seen the footage. Where there is the city, there is catastrophe, or risk of catastrophe, or certainty about a coming catastrophe. There is also, though, great pleasure and play, sometimes even enmeshed with a vision of the city's dark futures, or its precarious present. The relationships between catastrophe and play in the 21st-century city also occupy my explorations here.

As I began to think through these relationships, I found that much of the compelling work being done by urban activists, artists, and scholars share a common interest in challenging one-directional spatial forms and metaphors. They complicate notions of perspective and point-of-view and instead seek what Stephen Graham has called "the vertical and volumetric geographies of our world."[2] To grasp the city and its perils and pleasures requires not only moving across the skyline or the city street but also upward and outward.

In the early 2000s, I was doing field research in Buenos Aires with a group of artists, activists, and technophiles working under the banner of Buenos Aires Libre (the group and the autonomous network they built are the focus of the case study presented in Chapter 2). As part of that research, I climbed a rickety ladder to the rooftop of an office building in the *Once* neighborhood of the city where the group was installing a directional satellite dish to form a node in their network. The view of the city from above, the way in which the group sent their signals through the air to meet other dishes installed on other rooftops by other members across Buenos Aires, and the sheer pleasure of climbing itself, all indicated that playing in the city can and does take place on vertical trajectories, and more important, that political contestation can and does as well. Part of the logic behind building this autonomous network was to circumvent the policing of information on the Internet, and to perform the hopeful work that ensured that if catastrophe destroyed the formal network, Buenos Aires Libre's own homemade one could sprawl across the city, ready to more sustainably and equitably take its place. Playing with and refiguring the vertical by sending signals from rooftop to rooftop, by revaluing the horizontal through a charging of sprawl as a site of communal informational exchange, allowed Buenos Aires Libre to build a richer view of the city. Their vision of catastrophe was, too, a part of their contestation of certain vertical forms of power.

The other case studies in this book affirm the possibilities afforded by a refiguration of and play with the vertical. In Miami, the graffiti artist (or artists) who scaled the abandoned hospital building in South Beach to write in enormous, black letters the following sentence: "YOUR MILLION DOLLAR HOUSES WILL SOON BE UNDER WATER," (hereafter, YOUR, Chapter 3) likely found some pleasure in the precarious climb and the bold trace she left, anonymously, at the top of the building. The work also points toward another vertical arrangement that might yet be altered: class stratification, in which those with substantial capital are positioned, sometimes literally, above those without. And, of course, the graffiti piece points to the slowly shifting encroachment on the city's borders (both those underground at the water table and those at the city's coasts) of warming saltwater. Pigeonblog, a participatory lay science project designed by Beatriz da Costa and both human and non-human collaborators, used carrier pigeons to monitor pollution in Los Angeles and Southern California (treated in Chapter 4). The project clearly understood the fun that could be had in the air, even as it located the vertical field as the seat of certain forms of surveillance, control, and biological risk. Da Costa, whom I interviewed shortly after the project was completed, discussed both cell towers and bird's-eye views as essential to both her unruly intervention and to some of the forms of power against which it was aimed. That the project piggybacked on networked telecommunications meant it reached both into the air and into the ground below, leveraging SMS technology, on the one hand, and terrestrial Internet infrastructure on the other.

Lisa Parks' recent work, *Rethinking Media Coverage: Vertical Mediations and the War on Terror*, informs my own here. Parks argues that the vertical field has become valuable terrain through which power moves. "Vertical hegemony," she writes, is

> the ongoing struggle for dominance or control over the *vertical field*, which here includes combinations of terrestrial, aerial, spectral, and/or orbital domains. [...It] cannot simply be achieved 'out there' in open skies. To register or take effect, it must be communicated through and materialized as part of culture on earth."[3]

My intent is to answer her call to "foreground the constructedness of vertical forms of power,"[4,5] by examining the ways in which aesthetic and material interventions in the city, like those performed by the projects discussed in this book, are already working on upending and reforming the vertical in novel and resistant ways.

Both Parks and Graham have made great contributions to thinking through verticality as both a problematic and material phenomena essential to understanding power. Graham's work is an exhaustive read of verticality in the city. Parks' of the nature of vertical power and mediation in the post-9/11 era. Both point to some full-fledged catastrophes attached to verticality. But neither have been invested in the kinds of aesthetic interventions this book explores. It is my contention that such interventions offer not only complex models of the city itself, and of contemporary forces of surveillance and control within and through it, but also leverage their play toward novel forms of being in and reshaping the city.

On methodology

In exploring the three works that are the case studies of this book in their specific urban contexts, my methodology has been in part ethnographic. I conducted interviews with participants and creators, attended meetings and workshops, wandered through website as a user, and participated in events hosted by the artists and activists behind these objects and practices. I lived and worked in all three cities during the research and writing process and thus have, like all city dwellers, been a material participant in the production of these urban ecologies. Pigeonblog, Buenos Aires Libre, and YOUR are invested in the relationships between aesthetics, objects, places, and viewer or user practices and experiences in the cities they explore, and part of the practice of this book is to plug in, so to speak, to these relationships as one means of understanding them.

Nicolas Bourriaud's writing in *Relational Aesthetics* argues that works of art that engage relationships between and among people and things, creating linkages between rather than maintaining distances, have the capacity to produce critical relationships to the worlds we inhabit. "The reigning ideology would have the artist be a loner, imagining him solitary and irredentist [...]," he writes.

> This rather naive imagery muddles two quite different notions: the artist's refusal of the communal rules currently in force, and the refusal of the collective. If we must reject all manner of imposed communalism, it is precisely to replace it by invented relational networks."[6]

The objects of this study are all aiming at the production of new "relational networks." Their play in the vertical field helps them do so. In my view, it is very difficult to critically understand such works without entering into the new networks they produce and imagine. That is, after all, precisely what they are asking viewers and participants to do and how they manage to work, particularly when they so publicly move within the space of the city.

I have supplemented my participatory research with work and writing more typical of humanistic studies. In the case of YOUR, this was the only possible route as the artist or artists responsible for it remain anonymous. These urban objects and practices are texts to be read closely, and in doing so, I look not only at the infrastructures, cultural and political environments, ecologies, technologies, and economies that help bring them into being and into which they feedback, but also have sought to understand where they fit in a trajectory of urban media more broadly. Part of the project of this book is, thus, also a genealogical one. I take seriously the critique made by, among others, Parks in her work on residual media, "Falling Apart: Electronics Salvaging and the Global Media Economy."[7] "New and Old" media are relative terms. Setting them up in an oppositional relationship to each other, particularly when examining emerging practices that use so-called new technologies, risks reinforcing the accumulative and progressive logics of capital and reproducing many of the ideological veils built to hide both the material histories of technologies and urban media(ted) practices as well as their potentially damaging futures.

Though the concerns this book hopes to address are many, its overriding thesis is that playing with and in the vertical field, even and perhaps especially to address ongoing, looming, and even imagined catastrophes, can and does reshape cities. These works show us the kinds of worlds we inhabit in the city, forces that govern it, and risks we face. Ultimately, their interventions point toward alternative futures in which the common and the collective can gain a stronger foothold in the city, exposing the constructedness of vertical power and opening up ways of making sustainable and equitable urban worlds. They leverage the catastrophes they point to and imagine for revolutionary hope. Sometimes imagining or refiguring catastrophe, playing in the city while speculating on its undoing, is how we strive to change its shape so that it better bends to our desires, connections, and (human and non-human) bodies. Imagining catastrophe is constitutive with *thinking* the city at all these days and it may be a necessary step toward both surviving urban catastrophe and changing the sometimes disastrous shapes of the cities of the present.

Thinking in terms of catastrophe is found throughout much academic thought. Scholars in media, urban, and cultural studies and other disciplines have entertained, imagined, and understood cities as bound up with it. Catastrophic vision is equally visible in contemporary global capitalism. Whole industries are devoted to translating coming or current catastrophes into wealth for those who own the means of production. Economic precarity, palpable on many urban stages, is its own ongoing catastrophe, one that has been articulated particularly well by David Harvey in a number of his works on the urban, and recently and elegantly by Lauren Berlant, and others who will appear as my interlocutors in the following pages.

This book would not be possible without the work of Rob Nixon. His *Slow Violence and the Environmentalism of the Poor* explains so clearly that we must address the "incremental and accretive" violence that he calls "slow violence".[8] He is direct in calling this violence catastrophic, at scales that range from the personal to the planetary. The case studies in this book seek to understand the consequences of the catastrophe of slow violence within the specific contexts of the cities into which they intervene. Like Nixon, these projects understand the deep entanglement of the environment (and environmental catastrophe) and political, social, and representational structures. And they all make an attempt to do what Nixon seeks in environmental activism: they "plot and give figurative shape to the formless threats whose fatal repercussions are dispersed across space and time."[9]

The projects this book explores produce novel figurative shapes in and of the city in ways that respond to Nixon's call. In so doing they also reconfigure, rethink, and resist one-dimensional models; in fact, this kind of work helps to unearth the city's dynamism and the way power moves and is experienced with in it. Graham has convincingly argued that to understand the city we need "fully three-demensional and critical perspectives."[10] He notes that language about power and control are persistently couched in vertical terms: "vertical metaphors saturate – indeed, *constitute* – language about power, wealth, status and happiness. [...]: they both derive directly from the physical and phenomenological experience of social life and actively influence how people perceive and shape the social and political world."[11] Thus working to expose, to repurpose, and to shape cities anew means contending with verticality on both material and discursive fronts.

Verticality and catastrophe in the archive

Engagement with verticality is everywhere in the larger archive of theoretical works from which the current writing draws. Walter Benjamin's insights on the changing shape of the city under the yoke of industrial capitalism, his sense of the ways the urban world could be both a site of immense joy and delight and of constriction and habituation, produce views that are both local and expansive. His treatment of architecture and cinema allow his work to think in ways that disturb the tendency in urban studies and geography to read the city across a single horizontal plane at the expense of other directionalities. Henri Lefebvre's insights about the real social and political import of urban life, his "right to the city" and the many movements it has helped to spawn, certainly offer a more three-dimensional approach to urban thinking. Michel de Certeau's reading of the city as a space of articulation, of affordances and prohibitions employed, disregarded, or resisted by urban inhabitants and wanderers of all sorts pays close attention to the experience of perspective, from the pedestrian at ground level to the God's-eye view of the city planner. Harvey's analysis of the ways in which the urban site is central to the management of surplus, and to the workings of capitalism itself, understands the movement of capital and of resistance to its extraction of value in the city as multidirectional. More recent works such as Benjamin Bratton's *The Stack* and Keller Easterling's *Extrastatecraft* understand the city as distinctly informational, pinpointing (albeit differently) the ways in which its organizing structures are so often vertical. These are, broadly, the lines of thinking that all position the contemporary city as an arena essential to current workings of power, with its regimes of control, regulation, and surveillance. They are also all invitations to think beyond the horizon or the ground level. When juxtaposed, they can expose the city as a site rife with potential for activism, for pleasure, for alternative cultural and social productions and formations.

The word "catastrophe" in fact, is etymologically linked to vertical movement, as Eva Horn points out. It stems from the Greek preposition for "down" or "downward" (*kata*) and the verb "a turning" (*strephein*): "literally, catastrophe thus denotes 'a sudden downward turn.'"[12] And many of the thinkers that have approached verticality have had to contend with catastrophe as well. Sometimes catastrophe appears explicitly (as in Nixon's work, or Harvey's), sometimes it remains lurking within the function of state power, terrorism, extreme inequality, structural violence (as in the work of de Certeau, Lefebvre, Benjamin, and others). But there has emerged across a range of contemporary thought, too, an increasing sense of the catastrophic as something of an existential and affective reality in the present in ways it was not in the past. Horn's work (treated more expansively in Chapter 3) argues that the modern imagination has a symptomatic collective fantasy of the future *as* catastrophe. Berlant's *Cruel Optimism* (also discussed in Chapter 3) also locates in the experience of the present something that could be described as catastrophe: "the present moment increasingly imposes itself on consciousness as a moment in extended crisis."[13] Cruel optimism itself is a kind of catastrophe that is at once both personal and communal.

To think through these kinds of formulations in and through urban terrain is necessary because, in part, the city is where most of us live. According to one United Nations study, by 2050 nearly 70% of the world's population will live in urban areas.[14] Urbanization at a global scale has caused a few of its own economic, ecological, and social catastrophes, of course, laid out compellingly by Harvey in his work on neoliberal restructuring and the city, and by Mike Davis in his *Planet of Slums*.[15] The risks that urban populations face are only increasing.

The city is also the place where we will probably see catastrophe unfold, largely because of the media's (vertical) privileging of the urban environment. The rapid development and widespread use of digital technologies of information circulation contribute to the expansion of this privilege. Images of the city and discourse about it, particularly as a populous landscape in peril, circulate on a global scale in ways they have not before. For readers in the United States, the terrorist attacks on the World Trade Center in New York City in 2001 are perhaps the most exemplary. But there have been countless catastrophes circulated and looped on screens large and small before and since then: footage of floods, earthquakes, police violence, riots, and war, and such images contribute to a common sense of the catastrophic concentrated in distinct ways in the urban world. Western viewers seem deeply if morbidly attracted to these images of the city. Horn notes this in the opening pages of her book, locating in images of the deserted metropolis so many of our fantasies about the future as catastrophe. This was borne out recently with the explosion of photographs of abandoned urban centers across the world during the initial lockdowns of the coronavirus pandemic. It's worth noting, too, that these images were very often shot from above: yet again catastrophe is entangled with the vertical field.

Urban experiments, urban fantasies

The projects this book tackles respond to ecological, political, and economic challenges posed by the city in the contemporary Americas and they do so by reshaping the terrain they occupy as they speculate on the unraveling (political, economic, ecological) of that same terrain. They imagine and engage catastrophe as a mechanism to understand the city. They put a broad range of disciplines into conversation with each other – in art, critical theory, media studies, architecture, and urban studies, to name just a few. Comparative and interdisciplinary analysis is required to begin to read them because the meanings they make and manipulate touch on fields that are experiential and discursive, spatial and symbolic. They cross borders of all kinds.

These works are partial in their successes, often ambivalent in their political ambitions, and experimental in their investment in urban environments. They also occupy positions on the periphery of institutional practices in art, education, and industry (another way in which they resist certain vertical forms of power). This situates them in a particular space vis-a-vis the urban catastrophes they imagine, and in the cities in which these imaginings materialized. It allows them and their reader/critics to speculate on possible futures (not all disastrous futures) that might still come into being even when the projects themselves do not entirely picture them. In other words, because they are not yet wholly circumscribed within governmental, economic, or cultural institutional practices, the futures they imagine and the futures they allow the critic to imagine are unruly. In this way, they point toward alternative urban landscapes and alternative ways of making and playing within them that remain open, uncircumscribed. To return to Graham, they demand more vertical and volumetric geographies even as they subvert vertical power.

I treat the three cities of this exploration in separate chapters. I begin in Buenos Aires (Chapter 2), move on to Miami (Chapter 3), and then to Los Angeles (Chapter 4). Each of the projects I read also offers a distinct temporal framing for catastrophe and a specific model of city space. Pigeonblog's concern with nonhuman animal collaboration and pollution regulation points to ongoing catastrophe for the urban poor in Los Angeles and Southern California: catastrophe as unfolding, slowly, now. The urban landscape it envisions is thick. That is, the city moves in all directions: vertically, through the participation of avian collaborators, but also rhizomatically, through the informational networks and subterranean and terrestrial infrastructure they employ. Buenos Aires Libre's older network technologies, which might be used when the Internet is destroyed (by political or ecological crisis, or some other catastrophe yet to be envisioned), understand catastrophe as possible at any moment. Catastrophe is, for the network and the organization that builds it, opportunity. It too sees the city as both on the ground and in the air, but its primary activation of urban space is sprawl. YOUR's vision of catastrophe is teleological: it takes the doomsday form by proclaiming a nearing end. Its construction of urban space is of a malleable and material Miami: the city is plastic in its vision, in terms I will elaborate in detail in Chapter 3, and the forces that shape it are ecological, economic, and biological bodies (whether an artist, a king tide, or a condo tower).

All three works construct the experience of urban space as temporal, tactile, and mediated. All indicate an awareness that contemporary urban life as an extremely broad territory, both discursive and material. They are projects that are, regardless of political bent or scale of the endeavor, deeply concerned with urban life as it is now, and hopeful for how life, even in the age of urban catastrophe, could be.

The three case-study-oriented chapters are supplemented by three, short, contemplative interludes: Urban/sprawl, Plastic/city, and Messy/terrain. In each I outline the stakes of urban intervention now. Urban/sprawl advocates for "sprawl" as an epistemological approach to the city as well as a material phenomenon. It argues for a resuscitated radicality in sprawl, both in thinking on the city and in on-the-ground organization. Plastic/city makes an argument against neoliberal discourses of resilience and flexibility and argues for plasticity as the counter model for urban action and epistemology. Messy/terrain argues for the value of "messiness" and disorder in the urban milieu and tracks a history of advocations for mess in modern and contemporary urban theory.

These interludes are intended both as nodes that link the case studies in this work to a larger network of thinking on the cities of the Americas, situated as the intersection of technologies, ecologies, and layered apparatus of control (economic, political, social, "natural") and to offer a kind of theoretical scaffolding that will allow this work to invite interventions by other scholars, artists, activists, and city-dwellers of all kinds. They are also a way for me to play, like those whose work I read here, in and with the city.

The projects this book studies literally shape the city, if sometimes only briefly, as they imagine what it may be becoming. That is to say that the objects this book studies are speculative in nature. They imagine other possible cities as the play within their specific milieus and imagined or ongoing catastrophes. In this they align themselves with speculative fiction. Fredric Jameson reminds us that when we read speculative fiction critically,

> we must resist the reflex which concludes that the narrative fantasies which a collectivity entertains about its past and its future are 'merely' mythical, archetypal, and projective, as opposed to 'concepts' like progress or cyclical return,

> which can somehow be tested for their objective or even scientific validity. This reflex is itself the last symptom of that dissociation of the private and the public, the subject and the object, the personal and the political, which has characterized the social life of capitalism. A theory of some narrative *pensée sauvage*–what I have elsewhere termed the political unconscious–will, on the contrary, want to affirm the epistemological priority of such 'fantasy' in theory and praxis alike.[16]

Buenos Aires Libre, YOUR, and Pigeonblog are in some part fantasies of a future on its way. They are part of what is a collective narrative about both our urban pasts and our urban futures. They say something about delusions of progress and, like the speculative fiction in Jameson's reading, they also say something, if sometimes unintentionally, about history. This book also insists upon on the "epistemological priority" of the fantasies they have about the urban world,

> For Jameson the "deepest vocation" of speculative fiction is, contradictorily, to demonstrate and to dramatize our incapacity to imagine the future, to body forth, through apparently full representations which prove on closer inspection to be structurally and constitutively impoverished, the atrophy in our time of what Marcuse has called the *utopian imagination*, the imagination of otherness and radical difference; to succeed by failure, and to serve as unwitting and even unwilling vehicles for a mediation, which, setting forth for the unknown, finds itself irrevocably mired in the all-too-familiar, and thereby becomes unexpectedly transformed into a contemplation of our own absolute limits.[17]

The three imaginings of the future of the city that occupy my attention here also share, in many ways, "success by failure" in that they too set forth for the unknown, but remain mired in the familiar. They are not, whether they employ of-the-moment technologies and radical forms (Pigeonblog), or conventional methods of being in the city (YOUR), novel. They do not show us cities we cannot already, even easily, imagine. They are not the "apparently full representations" of Jameson's speculative fiction. When they imagine the city in ruins, the rubble is familiar. And like speculative fiction for Jameson, they end up telling us as much about where we are as where we might be going. Speculative though they may be, they are also speaking of the present. Lacking the rich textual tools of science fiction, their works of imagination are always partial and in this way, they offer a pragmatic way of imagining if not a future, then certainly a way of reshaping the city now.

These works envision catastrophe in the future to expose that catastrophe *is now.* In sometimes bizarre and circuitous ways, their relationship to urban disaster, sometimes explicit, sometimes gestural, provide a route to the recognition that humans are historical actors who *can and do act.* Albeit, of course, under (catastrophic) conditions they did not choose. In other words, they indicate agency in the urbanite, and they invite her to use it.

These works also very much position themselves against another speculative fiction that has of late blossomed among urban planners and policymakers. If there is an exemplary construct in the urban imagination of the illusion of a scientific "concept" like progress, it is the so-called smart city. Urban planners are heavily invested, both intellectually and financially, in this fantasty of the city. Adam Greenfield has succinctly unpacked the perils of this fantasy in his *Against the Smart City.* Buoyed, as Greenfield shows, by the corporate interests of firms like Cisco and Siemens, the "smart city" is a potent fantasy. He quotes Siemens corporate Web page: "Several decades from now, cities will have countless autonomous, intelligently functioning IT systems that will have perfect knowledge of users' habits and energy consumption, and provide optimum service."[18] Such a fantasy is not only straight out of science fiction, it imperils a vast continuum of life from personal privacy to national security. One issue that is increasingly pressing (in the city and beyond) is that the algorithms whose work would produce that "perfect knowledge" for the smart city are trained on the vertical hierarchies imposed by structural classism, racism, sexism, ableism, trans- and homophobia that undergird contemporary capitalism.[19]

The interventions explored in this book and the ways they envision the future city offer a striking counter to the capitalist fantasies of the smart city. Instead of "perfect knowledge of users' habits," Buenos Aires Libre relishes and uses urban disorder. Instead of order, YOUR intentionally produces disorder. Instead of striving for 'optimum' omniscience, Pigeonblog embraces incompleteness and messiness.

In the Messy/terrain interlude, I discuss the smart city in more detail. But it merits mention here because of the way it makes explicit the position of the city in the ideological struggles to which Jameson has devoted so much of his writing. It also points to the ways we need to rethink what urban technology is and could be, and the kind of collective narratives that are bound up with it.

City speculations

Much like the works and practices it explores, this book is itself speculative. For those of us who are still seeking modes of developing more equitable, just, and sustainable relationships among communities, animals, objects, and environments, speculation has distinct critical capacities. Here I speak of speculation as something broader than speculative fiction. Speculation's engagement with present political, biological, ecological, and economic structures, and its mapping of possible futures that pose alternative forms of knowing and producing the world make it an exceptional tool with which to examine and reimagine the urban. Speculation is concretely tied to perceptions and constructions of risk – financial, biological, social, ecological – and so can also serve to figure the political stakes of the fields at which it is directed. By outlining dangers, it can unearth new routes to mitigate them. By banking on change, it opens up space for challenge to and intervention in current urban structures.

Speculation is also one of the governing vectors of global capital. For many, the word speculation does not conjure fictions but figures; think futures trading, or the credit default swaps that helped trigger the 2008 financial crisis. Speculation's mutual imbrication with the current economic regime makes it a particularly slippery tool to use in imaginatively shaping a world in the dynamic process of coming into being, but a necessary one nonetheless. In this I follow the work of the anonymous collective, an uncertain commons, which has outlined two forms of speculation: firmative and affirmative. Firmative speculation aligns with vertical power. It

> renders latent possibilities as calculable outcomes : the regenerative qualities of a plant become measurable as medicinal capacity; the worker's embodied energy is formalized as specialized skills; creativity is reduced to intellectual property. Such translations into quantifiable capacity seek to harness and exploit potentiality, foreclosing other possibilities. We are most familiar with these forms of speculation, a predatory speculation that negates potentiality through a variety of mechanisms, turning open-ended futures into more of the same; it firms the status quo in the name of change.[20]

As should be particularly clear at the current historical juncture, this kind of speculation drastically and materially impacts the world. Population flows, urban blight and development, violent uprisings, austerity measures, political movements, and so on have been intimately tied to speculation in the urban landscape.

Many of the recent financial crises of the 2000s up to the present across the globe require a counter speculation that, instead of banking on debt or fear, deliriously reproducing and regulating life just as it breaks beyond its borders, bets instead on the capacity for imaginative production, for community formation, for sustainable, ecological urban practices. In the language of an uncertain commons, that counter speculation is affirmative, rather than firmative. It is speculation in which

> there is expectation, conjecture, and anticipation: modes of living that recognize the dormant energies of the quotidian and eventualities that escape the imagination. [...] To speculate affirmatively is to produce futures while refusing the foreclosure of potentialities, to hold on to the spectrum of possibilities while remaining open to multiple futures whose context of actualization can never be fully anticipated.[21]

Instead of foreclosing possible futures and limiting their scope under a regime that claims endless expansive capitalism as the only motor of production, the kind of speculation this book and the projects it studies hopes both to model and to locate in urban interventions in Buenos Aires, Miami, and Los Angeles, opens up multiple potential trajectories not circumscribed by the production of surplus. While financial speculation often covers over the concrete stuff of urban life and communication, speculation of the sort done by this writing and by the objects it studies have an especially potent power to unveil it. To imagine urban futures (after some catastrophe or not) in an attentive, critical way means, after all, to imagine building them – with cables and steel, plastics and plants, cybernetic bodies and all the actual stuff that makes up urban life and might be salvaged from it.

What both capitalist speculation and the kinds of urban imaginings done by artists, activists, and scholars share is an affinity for or relationship with catastrophe and verticality. It may seem contradictory to argue that these urban interventions offer alternative modes of understanding the urban landscape, and open up escape hatches in the global capital's regime even as they tend to foresee disaster and destruction. But, as Jameson notes, imagining the future (even if it is set in the ruins of the city) tells us something about where we are now. Capitalism employs speculation on disaster and works to impose vertical hegemony. Alternative forms of speculation on catastrophe, though, can offer counters to vertical hegemony even as they employ parallel relationships to the future.

One more note on speculation in urban terrain: Tung Hui-Hu has recently argued that infrastructure is "first and foremost a speculative medium" that "is best understood as a way of translating future capacity into the present."[22] He goes on to explain that part of its coming into being is about imagining itself as sustainable: "Infrastructure requires not just maintenance," Hui-Hu writes, "but also an imagination of its collapse so as to preempt and avoid that collapse; indeed, infrastructure is a way of designing the everyday to bear the load of emergency."[23] Well-engineered infrastructure is thus always speculatively oriented toward catastrophe. Cities, which are the sites of immense infrastructural investment, are thus always engaged in a certain catastrophic thinking.

The projects this book studies employ infrastructures in the vertical field in ways that run counter the fictions of vertical power. And in doing so, they imagine catastrophe. Over the course of this writing I argue that this play, at the intersections of the vertical field and urban catastrophe, can begin to formulate a politics of contestation that aims toward environmental justice in the broadest sense of the term. The work of this book is to begin to map the ways in which such a politics might already be at play in contemporary urban practices and in critical theoretical approaches to cities across the Americas, as well as ways in which such practices and theories can offer important openings into which urban practitioners might peer to refine and reform such politics in the future.

Urban (media) ecologies

Like all of the projects you will read about here, this book is concerned with the ecological, both in its concern with myriad forms of biological life and their various intersections (in the sense of the biosphere and the dynamic interaction of multiple forms of life and multiple objects) and in the way it outlines a number of conceptual territories that interact. The urban ecologies it imagines are closely allied with "media ecologies," as they are described by Matthew Fuller in his *Media Ecologies: Materialist Energies in Art and Technoculture*. Fuller asserts that ecology as a term "is one of the most expressive language currently has to indicate the massive and dynamic interrelation of processes and objects, beings and things, patterns and matter."[24] The term in Fuller and in this work maintains its slipperiness and dynamism. It cannot be exhaustively read through the vertical, alone. It requires exploration in multiple directions, across multiple scales. It morphs within each project explored and it remains plural because each intervention here tracked and contended with both interacts with and ushers into being different kinds of relationships. Urban ecologies, which are and have always also been media ecologies, exist as multiple, though never discrete.[25] They are never total, never whole. Modulation and porosity are key to understanding how they both come into being and continue to change. Their plasticity[26] is also central to the potential they offer the urban practitioner for change.

While Fuller's work leans heavily on the "media" of his media ecologies, and was not concerned with the ecological impacts of media per se (only very recently has there been a turn in media studies toward an ecocritical approach, as seen in Parks, Nicole Starosielski, Hui-Hu and others), I find his thinking useful in centralizing mediation in the urban interventions I explore. Media ecologies are insistent on space and place. Fuller is quite right to attend to material impacts and environments and the purpose of the analysis throughout my work here is to answer in some ways his call by distinctly locating, in three specific American cities, the workings of mediated ecological relationships. The urban is an exceptional site in which to unpack such relationships because no city can be reduced to ecology alone, nor to technology alone, however broadly or rigidly one might like to define these two terms.

The contribution of this writing, in its exploration of three specific cities, is to offer a hemispheric intervention. Such an approach is central in investigating the specificities of these three immigrant, global, port cities. The flows of information, bodies, and things between and through the Americas are brought distinctly into relief in all three. What's more, those flows have drastic and deep impact on the material, affective, and social structures in the urban landscape. Miami is home to the biggest Argentine community in the United States. Los Angeles and Miami are two of the three cities (alongside Manhattan) in the United States that have seen skyrocketing foreign investment in luxury real estate.[27] These are only two examples of the kinds of relationships between the urban environments of the Americas that demand both a local focus, and hemispheric awareness. Site specificity matters, as do links between sites.

Demonstration in the city (or, why city space still matters)

The urban texture to which the works explored herein contribute, as I have detailed above, is linked to catastrophes produced through the intertwining of anthropogenic climate change and contemporary capitalism. The latter's most recent and visible explosion was the global financial crisis of 2008. Harvey has described it as "undoubtedly, the mother of all crises."[28] He goes on:

> Yet it must also be seen as the culmination of a pattern of financial crises that had become both more frequent and deeper over the years since the last big crisis of capitalism in the 1970s and early 1980s. The financial crisis that rocked east and south-east Asia in 1997-8 was huge and spin-offs into Russia (which defaulted on its debt in 1998) and then Argentina in 2001 (precipitating a total collapse that led to political instability, factory occupations and take-overs, spontaneous highway blockades and the formation of neighbourhood collectives) were local catastrophes.[29]

These crises, whether they reverberated locally or globally, were, Harvey shows, very frequently "property- or urban-development led."[30] And while the abstractions of financialization may seem always to hover forever in the ether, the consequences of financial crises are absolutely felt on the ground, drastically altering the physical environment and experience of urban living across the Americas and globally. Harvey's take on these series of crises, *The Enigma of Capital*, begins, in fact, with a description of the havoc visibly wrought in U.S. cities after the 2008 collapse. In Cleveland, he writes,

> it looked like a 'financial Katrina' had hit the city. Abandoned and boarded-up houses dominated the landscape in poor, mainly black neighbourhoods. In California, the streets of whole towns, like Stockton, were likewise lined with empty and abandoned houses, while in Florida and Las Vegas condominiums stood empty. Those who had been foreclosed upon had to find accommodation elsewhere: tent cities began to form in California and Florida. Elsewhere, families either doubled up with friends and relatives or turned cramped motel rooms into instant homes.[31]

Harvey's reference to 2006's devastating Hurricane Katrina to describe conditions in cities post-economic crash is particularly evocative, but it's more than descriptive. Capitalism also has consequences for the way "natural" catastrophes impact the urban landscape. Those who most visibly suffered after the levees broke in New Orleans were poor, Black city-dwellers whose survival was as much tied to their position vis-a-vis capital, and the structural racism it employs, as much as it was to the fact that their city lay within the storm's path. That storm was empowered by the warmer ocean waters whose very heating is at least partially (to state it cautiously) a consequence of the industrialization of the modern world – a process that was fueled by the very accumulative logics of capital that are the root causes of the perpetual economic crises to which Harvey points. And, again, the valuation of some bodies and lives as valuable and some as disposable is a vertically minded, and catastrophic, structure.

The impacts of climate change, like those of financial crises, are interlinked and ongoing in cities globally: increasingly powerful storm systems worsened by warmer ocean waters, and flooding associated both with these storms and rising sea levels; drought-fueled wildfires; heatwaves; extreme temperatures. As the Intergovernmental Panel on Climate Change has concluded, and any number of recent urban disasters make obvious: cities are literally changing shape, and also changing policy, and changing social life across the globe in response to the heating planet.[32] To the images of a post-Katrina New Orleans that David Harvey employs we could add many more: stranded Houstonians using boats to cross submerged highways after Hurricane Harvey; San Juan neighborhoods transformed into rubble and an entire island in darkness after Hurricane Maria; flames leaping across Los Angeles' 405 freeway during the series of catastrophic wildfires that brought Southern California to its knees in 2017.

The catastrophic impacts of financial crises and climate change have incited responses from progressive groups (the Occupy movement in the United States and elsewhere, grass-roots, horizontal and communal actions in Argentina, environmentalist protests globally) that are counter-vertical, and distinctly spatial. Protestors, artists, and urban practitioners of all sorts continue to see the urban field as the appropriate place for vocal and visible resistance against the workings of multinational corporations, international banking institutions, and global capitalism more broadly, and against the removal of environmental protections or the full-scale ecological destruction wrought by both private and public actors (processes bound up with capitalist economic logic). Cities serve as both material landscapes in which to collect bodies in large gatherings and as sites that can be interacted with and manipulated in terms of their cultural signifying force. Indeed, Harvey argues in *Rebel Cities* that urbanity is not merely coincidental with resistant political movements, particularly anti-capitalist movements, but may be fundamental to them.[33]

Many of the practices developed in the Occupy movement, the global activist response to the 2008 crisis, and in the response to Argentina's 2001 financial collapse have continued into the present – as have some of the means of regulating and policing urban action. In Los Angeles specifically, Occupy Los Angeles or OLA continued to host general assembly meetings on a weekly basis until at least 2013, open to any who cared to join. Having been removed from City Hall, participants gathered, albeit in much smaller numbers, in Pershing Square (a suitable urban site given its centrality to the financial heart of downtown Los Angeles – perhaps even more so than City Hall) for months. In Argentina a practice called *cacerolazo* was utilized in the protests following the country's financial collapse. Participants at the Plaza de Mayo and elsewhere in the city banged on household pots and pans to make an incredible racket in public spaces. The practice is a favored form of protest across Latin America and has been more recently used by student protestors in Quebec and Chile.[34] In Florida, action has been varied and full-scale urban protests are rare, but much of Miami's urban art responds to the city's imperiled position between the Atlantic and the Gulf of Mexico: from murals in the arts-friendly Wynwood neighborhood to the illicit "YOUR" graffiti, the city's surfaces are visibly contending with environmental catastrophe: both imminent and already present.

It would be easy to look at the tactics of resistors in any of these sites as simply following in a tradition of protest in public space that has been ongoing, though wavering in intensity and duration, since at least the 1960s. And many of the demonstrations ongoing across the Americas are linked to that particular history. But the economic situation to which these manifestations of resistance are responding is not the same as the one against which general strikers of May 1968 rallied. For one thing, the stretch and shape of global links between states and financial institutions, between policy and environment, are different. Today's protests also have a certain valence in the present moment that they might not have had in earlier epochs, in part because of the persistent popular ideology that claims the local has disappeared, that place-based actions no longer carry weight when the real force of global capital is informational rather than spatial. (Paul Virilio has famously and frequently suggested that instantaneity of communication and the global vision it supposedly enables have quashed what legitimately local phenomena remained).

The projects described here suggest otherwise. They should be thought of as their own form of spatial occupation in the service of certain modes of resistance. They are not protests, of course, but they are demonstrations in the sense that they perform the alternative urban relationships they hope to expand.

The differences between Los Angeles, Miami, and Buenos Aires are also important in the specificities of urban resistance as well. Los Angeles has long been conceived as a city without a center, a sprawl in which public space cannot compete with the capsular space of the automobile, shuttling its riders along the network of Los Angeles freeways. Pigeonblog and other interventions suggest, however, that this vision of the city underestimates the force of space itself as central to relational urban experiences. Such projects work against a narrative of Los Angeles that is as atomizing and distancing as it is sprawling or, at the very least, point to ways in which space can be resuscitated as a site for play and protest.

In Buenos Aires, the occupation of space and work done in and on it carries an additional political weight. The particular pull of visibility and propinquity in urban terrain remains tied to the violent legacy of the military junta of 1976–1983 whose regulation of the political body was enacted as an erasure of bodies in space. The junta disappeared thousands of people during this era; silencing them and making them invisible. In its insistence on amplified sound and crowds protesting in public space, contemporary political protest in the city seems very aware of this legacy.

Miami's exaggerated porosity, both in terms of population flows and ecological shifts (hurricanes and tropical storms, warming waters along its coasts, rising sea level) render it distinct. Of the three cities, it is among the least studied, but this is changing as it is increasingly acknowledged as among the urban centers at greatest risk in the age of anthropogenic climate change, and as uniquely emblematic in terms of global population flows.

Putting contemporary concerns with space and place in Buenos Aires, Miami, and Los Angeles in conversation with each other helps highlight the many ways in which space itself can be read and engaged in urban practice. The intention here is not to conflate these three distinct cities, but to highlight the diversity of approaches to urban terrain, and to underscore the fact that across the Americas, any legitimate arguments about politics, economics, and ecology must take into account the very real need people still feel to be seen and heard in the city.

Buenos Aires/Miami/Los Angeles

There are any number of reasons why Buenos Aires, Los Angeles, and Miami should be put into conversation with each other. All are sprawling American cities, all immigrant cities, all highly developed and wired cities. I am foregoing a catalog of their cultural, infrastructural, political, or economic links because all of the theories of urban space and life referred to throughout this book understand that no city functions in isolation.

More specifically, I have also chosen Buenos Aires, Los Angeles, and Miami because while they are "global cities" in that they are centers of the global financial flows described by Saskia Sassen, they also speak also to a more hemispheric and geographically distinctive set of relations.[35] They can be seen as a way to triangulate the developed Americas, though as I read them, they are in no ways the only possible triangle to draw. The relationships set up in this book between the three cities should be read, then, as conversational in nature. My approach is intentional, that is, I want this book to employ some form of affirmative speculation as well. Part of the underlying analysis of urban media ecologies is the suggestion that juxtaposition has its own critical capacities, that it can conjecture without foreclosing possible futures. Positioning these three urban landscapes next to each other is meant to perform on the page what all of the projects and organizations I examine already begin to indicate in urban space.

Several theorists of these three cities from which the current project is drawn should help to flesh out how to think through them together. In Buenos Aires, Beatriz Sarlo's work offers a kind of prototype of a media-ecological approach as a critical rubric. *La Imaginación Téchnica* is an exceptional analysis of literary imaginations of science and technology in Argentina in the 20th century that does much to begin to map the links between narrative work, political environments, and the construction of productive epistemologies.[36] However, Sarlo is not concerned with the ways that the authors she writes about understood the interactions among nature, technology, and the urban terrain, which was increasingly the site of speculation in the fiction and journalism she reads. Her focus on the outsider and the underdog in the construction of the modern era in Argentina brings into focus what is at stake in the 20th-century scientific imagination, but it does not trace the feedback loop that technological innovation and shifting environmental realities have with the urban (both imagined and experienced).

There has been much research on Los Angeles and its relationship to the natural world. This may be due in large part to the work of Mike Davis in both *City of Quartz* and *Ecology of Fear*, which examine in great detail some of the more brutal ways the city has understood itself not as an ecology but as a metropolis capable of keeping ecological forces – most famously, perhaps, the Los Angeles river – at bay. Davis, though comprehensive in his reading of Los Angeles history, culture, and even ecology does not fully explore or elaborate the many ways that the city, and indeed all cities, still offer up places to play – places for pleasure not wholly circumscribed by consumer culture or governmental surveillance and control. Putting play more squarely into the picture I draw of contemporary Los Angeles will, I hope, serve as an invitation for further intervention from artists, critics, and urbanites of all sorts.

Miami has been treated in only a limited way by cultural scholars. There is a great deal of work on immigration and the city, for example, but only in cultural geography and urban studies are there more comprehensive looks at Miami. I will draw specifically on Alejandro Portes and Ariel C. Armory's work in *The Global Edge*, and on Jan Nijman's *Miami: Mistress of the Americas*. These works, though, are not concerned with single interventions or aesthetic practices outside of architecture. Aesthetics, in my writing, remain central. Pleasure may be something of a carrot at the end of my speculative stick, but it is an exceptionally useful urban tool and, as the following chapters will show, it also carries with it immense political potential.

Finally, of course, the job of any good comparatist is to look at texts in relation to one another, across national, temporal, and disciplinary boundaries. Buenos Aires, Miami, and Los Angeles can speak to each other and to scholars about urban life more effectively if they are put in conversation. They are American cities, one Northern, one Southern, one oddly both and neither, whose position in the cultural imagination of their continents looms very large; their weighty position beckons the critic to take a closer look.

Urban landscapes and escapes

The legal grey area of BAL, the collaborative co-production of Pigeonblog, the afterlife of the YOUR graffiti, which continued to circulate after it vanished from the building on which it was painted – all are devoted to imagining a city that is an escape route – a future city that may still be possible. That, of course, is the speculative goal of many creative urban imaginings: an ushering into being of a city that might work against the awful destruction wrought by capitalism's global regime on our natural, our political, and our social landscapes.

If these forms of protest in Buenos Aires, Miami, and Los Angeles to which I pointed earlier are any indication, play and protest and sometimes play *as* protest in urban space still have real discursive and political potential. Space, it seems, *still* matters, still must be played with, on, in, around. Perhaps if we can construct a strong enough vision of the space of the city as a space of play we might have a better chance at making it into legitimately *common* space. Ultimately, it is toward the common that this book and the projects it examines are aimed. It is my hope and belief that the city might serve as the site of the coming into being of the common, in the face of, or in preemption of catastrophe. It is, after all, a means by which to counter the logics of vertical power: there may be vertical vectors in the common, but never are there immobile vertical hierarchies.

This book is also intended to model a way of reading the city and in so doing to help open up space to remake the city. Each of the three projects offers an alternative paradigm of the city and thus a different way of exploring, thinking, and intervening in urban space. The readings here are not intended to be exhaustive; rather they are to be understood as starting points for further play. Urban interventions as objects of study can offer the critic a chance to speculate on her own future city and the practices in/on/through space that might someday shape it. Mediated urban interventions, indeed all imagining of the speculative sort this book advocates, can and do help form new and alternative relationships between ecologies, bodies, and technologies. As Manuell Castells once claimed: another city is possible.[37]

Notes

1 Joan Didion, *Slouching Toward Bethlehem* (New York: Penguin Books, 1979), 220-221.
2 Stephen Graham, *Vertical: The City from Satellites to Bunkers* (London: Verso, 2018), 15.
3 Lisa Parks, *Rethinking Media Coverage: Vertical Mediation and the War on Terror* (New York: Routledge, 2018), 4.

4 Parks, *Rethinking Media Coverage,* 3.
5 Parks, *Rethinking Media Coverage,* 4.
6 Nicolas Bourriaud, *Relational Aesthetics* (Dijon: Les Presses du Réel, 2002), 81.
7 Lisa Parks, "Falling Apart: Electronics Salvaging and the Global Media Economy," in *Residual Media,* ed. Charles Acland (Minneapolis: University of Minnesota Press, 2007), 32–47.
8 Rob Nixon, *Slow Violence and the Environmentalism of the Poor* (Cambridge: Harvard University Press, 2011), 2.
9 Nixon, *Slow Violence,* 10.
10 Graham, *Vertical,* 22.
11 Graham, *Vertical,* 15.
12 Eva Horn, *The Future as Catastrophe: Imagining Disaster in the Modern Age* (New York: Columbia University Press, 2018), 8.
13 Lauren Berlant, *Cruel Optimism* (Durham: Duke University Press, 2011), 7.
14 "68% of the World Population Projected to Live in Urban Areas by 2050, Says UN," *United Nations: Department of Economic and Social Affairs* (blog), May 16, 2018, https://www.un.org/development/desa/en/news/population/2018-revision-of-world-urbanization-prospects.html.
15 Mike Davis, *Planet of Slums* (London and New York: Verso, 2006).
16 Fredric Jameson, "Progress vs. Utopia: Or, Can We Imagine the Future" *Science Fiction Studies* Vol. 9, No. 2, (July, 1982), 147-148.
17 Jameson, *Progress,* 153.
18 Adam Greenfield, *Against the Smart City* (New York: Do projects. 1.3 edition, 2013). Electronic publication.
19 See, for example, Sofia Umoja Noble *Algorithms of Oppression* (New York: New York University Press, 2018). Also discussed in the "Messy/terrain" interlude of this book.
20 an uncertain commons, *Speculate This!* (Durham: Duke University Press, 2013). n.p.
21 an uncertain commons, *Speculate This!,* n.p.
22 Tung-Hui Hu, "Black Boxes and Green Lights: Media, Infrastructure, and the Future at Any Cost," *English Language Notes* 55, no. 1–2 (Fall 2017): 81–88. (82-83)
23 Hui-Hu, "Black Boxes," 83.
24 Matthew Fuller, *Media Ecologies: Materialist Energies in Art and Technoculture* (Cambridge: The MIT Press, 2005), 2-3.
25 See Shannon Mattern, *Code and Clay, Data and Dirt: 5000 Years of Urban Media* (Minneapolis: University of Minnesota Press, 2017).
26 I develop my insistent reliance on 'plasticity' over 'flexibility' or 'resilience' in Chapter 3 and in the Plastic/city interlude.
27 Rebekah Bell, "Foreign Investment in the U.S. Luxury Real Estate Market Reaches a New High," *Robb Report* (blog), February 2, 2018, https://robbreport.com/shelter/spaces/foreign-investments-miami-manhttan-los-angeles-2775258/.
28 Though as of this writing it is too early to say, the current economic fallout tied to the Coronavirus pandemic is likely to mimic in some ways and exceed in many others the 2008 crisis.
29 David Harvey, *The Enigma of Capital and the Crises of Capitalism* (New York: Oxford University Press, 2011), 6.
30 Harvey, *The Enigma of Capital,* 8.
31 Harvey, *The Enigma of Capital,* 2.
32 "Global Warming of 1.5°C. An IPCC Special Report on the Impacts of Global Warming of 1.5°C above Pre-Industrial Levels and Related Global Greenhouse Gas Emission Pathways, in the Context of Strengthening the Global Response to the Threat of Climate Change, Sustainable Development, and Efforts to Eradicate Poverty" (Intergovernmental Panel on Climate Change, 2018), https://www.ipcc.ch/sr15/chapter/spm/.
33 David Harvey, *Rebel Cities: From the Right to the City to the Urban Revolution* (London: Verso, 2012).

34 Johnathan Stern has written about the incredible experiential power of this form of protest. See Jonathan Sterne, "Quebec's #casseroles: On Participation, Percussion and Protest," *Sounding Out!* (blog), June 4, 2012, https://soundstudiesblog.com/?s=casserole.

35 Saskia Sassen. *The Global City: New York, London, Tokyo.* 2 ed. (Princeton: Princeton University Press, 2001).

36 Beatriz Sarlo. *La imaginación técnica* (Buenos Aires: Ediciones Nueva Visión, 1992).

37 Lauren Bon, "Another City Is Possible," *Metabolic Studio* (blog), n.d., https://www.metabolicstudio.org/64.

2 Alternative sprawls, junk cities

Buenos Aires Libre and horizontal urban knowledges

In 2010 and 2011, a group of artists, eco-activists, and programmers took over the top floor of a run-down government building in the working-class Once neighborhood of Buenos Aires. The space and its collection of art and community gardening workshops was known as Once Libre. Leaning against the wall of its defunct elevator shaft was a rickety metal ladder leading to the building's rooftop. A handful of wires and cables were threaded between two old PCs pushed up against the wall of the shaft and then strung up through it to an antenna, strapped to a tower built to reach far above the other structures in the surrounding neighborhood. These were the clunky materials that made Once Libre one node in the Buenos Aires Libre network. Buenos Aires Libre (hereafter BAL) is the focus of the this chapter. A horizontally organized group, BAL's ongoing project is the construction of an autonomous network, separate from the Internet, in and around Buenos Aires. Using wireless technology and line-of-sight connection between nodes, the BAL network is built for and by its users. Each node is constructed and maintained under the command of the members or participants who offer up spaces they have access to for the network, so the materials and manner of production vary. The nodes connect to each other (though not all connect) via electromagnetic waves. When I worked with the organization in 2011, BAL actively encouraged the use of found or recycled materials but some members preferred to build their nodes with entirely new, purchased equipment. Even then, the technologies the group uses are older and simpler than those on which the Internet depends. Though most of the nodes harvest electricity from the city's grid, the manner in which they connect is through the common space of the air above the city and not via fiber optic cables and copper wiring. (Each node can connect to those around it but, unlike other network technologies, there is no central node, no underlying base structure.) The network is entirely devoted to content sharing – music and video files, open-source code, images, and the like. The information available on it is also organized horizontally. No universal protocol structures it. Figure 2.1 illustrates some examples BAL posts on its wiki site of how to use the network.[1]

In examining BAL's work, I employ two key material spaces that double as metaphors: sprawl and junk. "Sprawl," as in urban sprawl, is both the material horizontality of expanding urban landscapes and the sprawl of BAL's expanding network over the city. Its double function here underscores the technological and

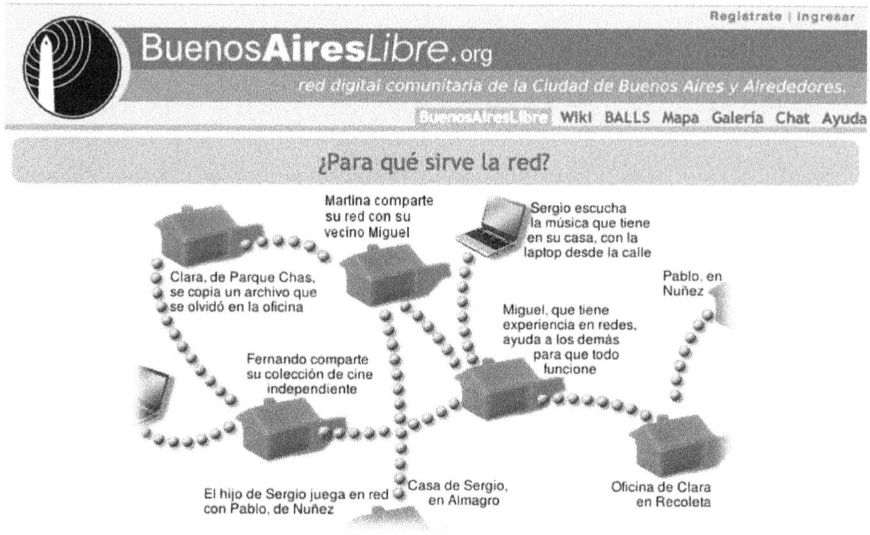

Figure 2.1 An illustration of possible uses for the BAL network
Source: Screenshot by the author.

organizational horizontality of BAL. "Junk" references both its material presence in the junked materials that BAL repurposes to construct and use the network and excess and redundancy situated in a transitional position, in a more politically and symbolically charged way. In this latter sense, the present chapter attempts to think through some of the work done and the spaces occupied by the BAL network: these are spaces that are made liminal, peripheral and other by political, informational, and economic systems in Buenos Aires specifically, but also globally. BAL recycles and reuses these in-between spaces and things in its attempts to remake and re-imagine urban communication. It is in its "junkiness," too, that BAL positions itself within the larger aesthetic urban movements to which this book is pointing: junk is the foundation of BAL's imagination of urban catastrophe, and the material it uses to work against vertical hierarchies and forms of power.

Both sprawl and junk, I argue, let us construct alternative approaches to thinking about urban life. Junk and sprawl are simultaneously part of and paradoxically covered over by certain governmental, infrastructural, and biopolitical systems – and, for that reason, constitute sites that have immense potential in the current moment, not only in the BAL network, but in urban landscapes understood more broadly. Sprawl and junk are also useful categories through which to consider the BAL project because they function to highlight the mutual imbrication of the informational and the material. BAL refigures the city of Buenos Aires – by setting up nodal towers it quite literally changes the landscape. While this is true of other information and communication infrastructures and technologies, part of the speculative potential of BAL is that it directly and explicitly imagines and shapes the city as part of its project. Perhaps more so than the Internet, the conditions of possibility for the expansion of the BAL network demand a knowledge of and an interaction with the environment on the part of the user. One of the goals of the exploration of BAL that follows is to unveil the centrality of space not just to the material presence of the city itself but also to its ideological, political, and informational structures.

Before examining BAL via junk and sprawl more specifically, it is worth finishing the story of the particular node in barrio Once with which I began this chapter. Throughout its brief run, Once Libre remained in something of a legal gray area. Participants more or less appropriated the space without having all of the requisite city permissions. The municipal government (encouraged, rumor had it, by a local labor union housed in the same building) eventually pushed Once Libre out of the space by cutting the padlock on the entrance gate and replacing it with their own in June of 2011. Legally ambiguous practices like those of Once Libre are typical among many, though certainly not all, of the members of BAL (several of whom are part of the hacker community). Among those in this camp, a commonly shared distrust of the government coupled with the challenges of navigating notoriously slow and notoriously corrupt municipal and federal bureaucracies encourages the belief that projects are more likely to succeed if permission is asked after the fact (this maxim was repeated to me several times by a founding member of BAL who goes by the handle "Vampi"). This position has benefits and drawbacks, but what is important to note here here is that liminality, a certain "outsider" status, is a starting point for BAL's interventionist practices. The group itself thus functions as a certain kind of junky presence, at least by the term's of the city's government and juridical structures.

Horizontalidad: Sprawling as praxis

Urban sprawl, the material horizontality of expanding urbanity (and sub-urbanity) is fast becoming a global norm. As Mike Davis writes, "sprawl has long ceased to be a distinctively North American phenomenon, if it ever was one."[2] While the largest expansion of Buenos Aires specifically occurred between 1945 and 1980, sprawl was not only a material fact of the city before and after that, but was also central to the debates that shaped it across the 20th century. Urban historian Adrián Gorelik and architect Graciela Silvestri write that "the entire modernization of Buenos Aires could be limited to the question of its urban sprawl: how to control, characterize, and define it, although the urban sprawl itself was unintended and never part of the project."[3] One-third of the total population of Argentina now live in the greater metropolitan area of Buenos Aires. According to the Argentine census bureau, as of 2010, 9,916,715 residents inhabit the 24 *partidos* – neighborhoods technically external to the federal district but whose boundaries with it are highly porous and practically invisible.[4]

In the Western popular imagination, Buenos Aires – sprawling though it certainly is – has managed to adhere to a more classical urban paradigm, largely because, having been planned and constructed to reproduce the image of modern Western European cities – it is often compared to Paris – it boasts a central urban core. As the country's capital, its primary commercial districts ring the Casa Rosada and the Plaza de Mayo, on the one hand, and the nearby federal courthouse and congressional building on the other. But any centrality one might hope to locate in El Centro, at the seat of the country's government, or in its financial district, is largely destabilized by the long and circuitous links the city has – in terms of both politics and infrastructure – to the vast and complex network of neighborhoods sprawling out from this urban core.

The sprawling nature of Buenos Aires makes the task of building an autonomous network very difficult, particularly given the requirements of BAL's technology.[5] The city is so large that without the aid of enormous material linkages, it would be nearly impossible to connect every person or even every neighborhood into the network. But the city's sprawl also means that the BAL network links its users in a way that the Internet does not. Each node is bound to the network by members who have to knowingly maintain their own position in the sprawl. They must physically direct the connections between their nodes by angling satellite dishes. The project requires a certain awareness of the network qua network. BAL thus depends on the fact of its locality – it is a network situated very specifically in the urban landscape of Buenos Aires – and also on a specific kind of knowledge about and in that locality.

Knowledge of the city's sprawl in terms of network connections and physical landscape is paired in the project with BAL's organizational horizontality, and it is here that the group most visibly works against vertically organized power. No single member of the group claims to lead it. No single participant functions as figurehead. Over the course of the months I spent attending BAL events and speaking with users and members, all were adamant that they did not speak for other members nor on behalf of the group itself. The monthly membership meetings are open to any interested party. During my time in Buenos Aires, these meetings were sparsely populated, usually with between five and twelve attendees. The BAL wiki maps, however, that track online nodes usually showed between 150 and 200 actively functioning around the city plus another 100 or more offline.

Nodes can be built by anyone, anywhere. Any participant who attends three meetings gains full membership privileges, meaning their vote in organizational decisions is equal to that of all other members. In addition to the monthly meetings, the group holds discussions and debates via a Listserv (to which anyone can subscribe and post) and the BAL wiki. All meetings are entirely open to visitors, and old and new members.

BAL's stated goal, to build a *red communitaria*/communal network is facilitated by its organizational model.[6] The point of the network is to share any information users care to upload to nodes with a broad community and to extend access as widely as possible across the urban landscape. This double horizontality, in terms of links between users and neighborhoods across space, is key to BAL's interventional potential and its material investment in the urban landscape. Such an organizational structure is typical of progressive groups throughout Argentina (though BAL claims no political affiliation or position). Birthed, in large part, from the movements that emerged to respond to Argentina's 2001/2002 catastrophic financial collapse, *horizontalidad* continues to be the salient organizational model for activists and workers.[7] During the collapse, this included cooperative takeovers by workers in businesses that were destroyed by the economic shutdown and as neighborhood organizations that made decisions about appropriate political action and resource sharing. Many of these organizations and co-ops are still operational, nearly 20 years later. One notable example is the publishing cooperative Eloísa Cartonera.[8] Launched in 2003, the cooperative grew out of the economic crisis and has a compelling relationship to the recycling economy of Buenos Aires. The covers of the handmade books they produce are made of cardboard purchased from the city's *cartoneros*, or trash pickers, who collect reusable materials in the city and its surroundings. Here we see another group for whom junk is a charged material with real political and economic value.

BAL must consistently confront the material fact of urban sprawl in Buenos Aires. Dedicated to widening access to users across the city and its surroundings, BAL frequently faces the problem of very large areas that are not yet linked into the network. Grappling with this gives members a knowledge of the space of the city, its breadth and depth, that the Internet cannot. Because BAL sees itself as a network built for and by its own users, there is also an understanding of the potential for a certain kind of communal social formation that is linked with this knowledge of the sprawl.

Unlike the mid 20th-century policy-makers and urban planners discussed by Gorelik and Silvestri, containing or controlling sprawl is not a concern for BAL. Rather, knowledge of the city's physical sprawl becomes a condition for understanding the expanse of the network infrastructure as well as the kinds of communities and users that infrastructure links: the more linked nodes, the better the network, the wider the variety of content available through it and the more diverse its users. It is not a matter of ordering the space of the city but rather of refusing the imagined boundaries between center and suburb in an attempt to link the broadest possible community of users over the largest expanse of urban landscape.

Among the few clear declarations of some kind of organization-wide ethos, BAL's home web page asserts: "valoramos la cooperación, tolerancia, innovación y solidaridad en nuestra comunidad"/"We value cooperation, tolerance, innovation and solidarity in our community."[9]

The group sees its work coming into being in space across the city and its outskirts as much as it does via the content sharing users can engage in online. They count their growth as much in nodes as gigabytes of shared information. Such community formation demands a horizontality that is spatial in order for the network to reach a broad base of users. BAL is not a network devoted to connecting immediately proximal users, but rather one that benefits from sprawling across a large body of land. Its organizational horizontality aids in this pursuit because it assures that users from very diverse neighborhoods throughout the unevenly developed urban sprawl are valued equally when they come to the table. Building a node in one of the city's wealthiest neighborhoods – Recoleta, for example – is no more important to the expanse and function of the network as building a node in one of the city's poorest – such as La Boca. Horizontality is thus activated by the group's materal and social practice, which it leverages against vertical hierarchies that might otherwise privilege certain members or certain spaces.

In addition to bringing into being a certain kind of community via the network, BAL also materially alters the city sprawl. The presence of the nodal towers changes the physical and informational landscape of Buenos Aires. The line-of-site requirement for the nodes encourages placing towers on the highest possible buildings in a given area: in building each node BAL members not only extend the network and shape the city, but gain access to new views of city space. This gives them a distinct picture of the materiality of information in the urban landscape. As they build nodes, repair them, visit them, or show them to new BAL participants, members and users see the city in ways others do not. They see infrastructure – including the network's own – and the organization of neighborhoods, their park spaces and concentrations of types of architecture. They not only see but spend considerable time working on the roofs of buildings, in elevator shafts, on fire escapes, etc. The vision of city space offered to them from alongside network antennas includes even some wildlife (birds, in particular, can be seen in different ways from above). And unlike the vantage point offered by the windowed skyscrapers of Buenos Aires, the relationship to the world above and below is distinctly haptic. To build a node, members climb, touch, and smell, with few barriers between their position and the immediate environment. These above-the-sprawl interactions with the picture of the city are not just a side benefit of BAL's work. They situate and contextualize participants and the network within the ecology of Buenos Aires. BAL leverages space in the vertical urban field to offer nuanced, material knowledge of, and broader informational access within, the city.

Marc Augé's work on "non-places" points to the way aerial shots and satellite photographs, taken from the vertical field, "habituate us to a global view of things."[10] Augé argues that this vantage point hides the ruptures and inequalities that globalization has increased (while the ideology of global connectivity manages to cover over these same ruptures and inequalities). The aerial views BAL offers, however, are not those of a humanist universalism, a frontierless global space where no one is excluded. The city seen from the node's antenna is a city unmediated by the screen. Weather systems, distances, the stability of one's footing – these variable phenomena mean that when building or visiting a node, users encounter the city as a physical, living entity in all its disorderliness. It is easy to see, too, the unevenness and asymmetry of urban development in Buenos Aires at such vantage points. While the Google satellite image of Buenos Aires in Figure 2.2 offers the urban as an ordered system, that is not how the city looks from its rooftops.[11] BAL engages a radically different understanding of city space via its interaction with the various levels of the city: it is a knowledge that is tactile as well as visual (see Figure 2.3).[12]

Remapping space, teaching sprawl

BAL asks users to understand the informational trajectories of the city in terms of localized bodies in space. In constructing the nodes, members have to know where in the urban landscape the other nodes are placed so they can direct antennas toward those other city spaces in order to connect to the larger network. BAL also runs a real-time map of nodes on its website, detailing which nodes are active and where, which nodes are under construction, which are offline and pointing out areas of the city that lack nodes or places where nodes exist but have yet to be connected to the network (see Figure 2.4).[13]

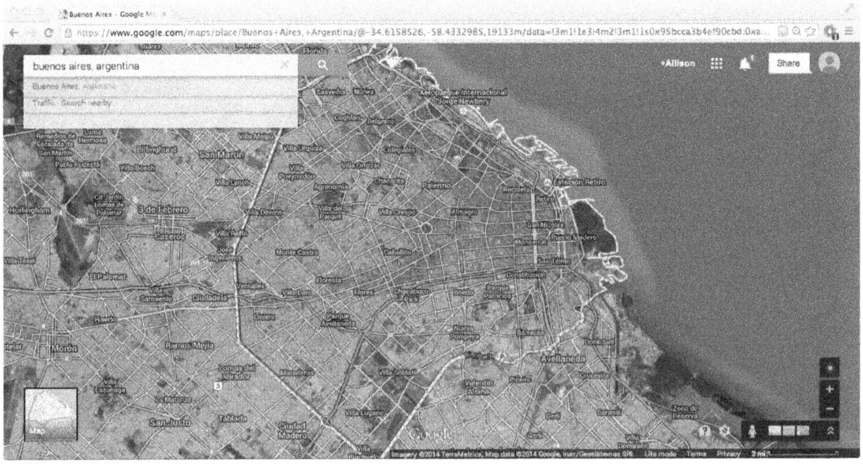

Figure 2.2 A Google map of Buenos Aires
Source: Screenshot by author.

Figure 2.3 Two members of BAL working on a node
Source: Photo courtesy Buenos Aires Libre.

Even users who do not themselves build or run a node have to understand the network in material, spatial terms to use it. Connections are not automatic. People who happen upon a node via wireless frequency will be linked to information about the specific node to which they are connecting and to information about the larger network. This information is always as spatial as it is "virtual." Each node is named by

Figure 2.4 A real-time map of nodes on the BAL website
Source: Screenshot by author.

the user who builds it; each node is mappable on the BAL site, so users come to understand the network in spatial terms. Even when this is not the case, when a user connects to the network the name of the node and the handle of its builder is made clear.

BAL educates new users and workshop attendees about the slow and clunky building process, thus working against conceptions of ephemerality and speed that tend to picture network technologies as access points to immaterial, spaceless, and bodiless exchange. These workshops are hands-on, inviting participants to literally touch and play with the materials that make the network. The connection of the network with the built environment of the city is hard to miss – in part because the BAL wiki includes numerous photos documenting the construction process. These images reinforce an understanding of the network's materiality.

The limited nature of the BAL network contributes to such a distinctly localized, material epistemology of the city. When a user pulls up the real-time map of nodes on the BAL website to examine network links, what she sees is an incomplete connectivity. Imagine what the ideological impact of such a map of the Internet would be. If part of the representation of global connectivity was also a documentation of its gaps, of the spaces unreached by fiber optic cables and telephone wires, it would be much more difficult to believe that an entirely inclusive, entirely global networked community existed. If such a representation of the network existed, users could literally picture the excluded spaces. Imagine, too, what the Internet would look like if average users were encouraged to trace the network's construction – if workers were named and cables were mapped. Users could, as those of BAL do, understand the network as the outcome of a process of material and spatial construction. They would be able to grasp at the real social nature of its coming into being, its maintenance, and its growth.

Theorizing urban practice and play in this distinctly material, spatial way can unearth possibilities for reshaping the city that already exist in urban space itself. Paul Virilio[14] and many others have done considerable writing on the swallowing of space by the compression of time in the contemporary moment, and they have linked this apparent phenomenon to developments in contemporary communication technologies. BAL fights this by reasserting the value of space and the material impact of network technologies. BAL highlights the sorts of present intersections (of forces both political and spatial) with which urban movements need to contend.

It matters that BAL is organized horizontally *and* is horizontally present in the sprawl. Any number of recent events throughout the world (the so-called Arab Spring, the Occupy Wall Street movement, the mass student protests in Chile and Canada, and the currently unfolding mass uprising against white supremacy), have shown that public and private spaces, material and architectural, remain very present in the exercise of power. One need only look at the recent marches and ongoing struggles around monuments linked to settler-colonialism and the enslavement of Black people, if this is in any real doubt. Sprawl needs to be re-examined in its relationship to horizontality and the political possibilities it may afford those who hope to build an alternative, common public sphere – one that is subject more to the whim of the people than to the fluctuations of global capital. Space has *not* vanished from the political landscape as a viable site of occupation and intervention, even if technologies have expanded our virtual reach. Radical play and practice as well as biopolitical control are dependent upon certain spatial configurations and certain ways of representing space.

Architect and theorist Keller Easterling has written compellingly on urban space's distinct political and informational force: "Spaces and urban arrangements are usually treated as collections of objects or volumes, not as actors. Yet the organization itself is active. It is doing something, and changes in the organization constitute information."[15] Her discussion of contemporary architectural practice clearly indicates the mutual imbrication of the material and the informational, indeed the material *as* informational. BAL's practice encourages an investment in the materiality of communication infrastructures as well as the prohibitions and affordances of the urban landscape on the movement of bodies (both human bodies and bodies of data). Power has always been invested in the spatial, even when its work is informational. To neglect its movements on and through space in favor of theories that understand it only in terms of virtual presence and quickening temporalities is to ignore the very material substrate on which power depends, and upon which its blows fall.

Junk network, junk city

The urban sprawl that is Buenos Aires is unevenly developed. A number of its neighborhoods are almost entirely without infrastructure. The *villas*, or slums, found in various locations throughout the city, and its populous outlying landscapes also function as liminal spaces in relation to the central presence of municipal power. They often lack running water and power except that which residents steal and jerry-rig. In one *villa* I visited, an independent, functioning pirate radio station hoped to activate a node in the BAL network to which it could link its broadcasts. Its electrical power was entirely pulled from some distant source by a long string of extension cords and adapters.

The diversity of spaces in Buenos Aires makes BAL a complicated project in relation to the economic realities of the sprawl. Nodes in the *villas*, at least, could serve to redress the problems of access and communication: it is toward this goal that most autonomous networks are built. In this sense, the work done by Rhizomatica to develop community-owned and community-operated autonomous cellular networks in rural Mexico and Nigeria offers a comparison.[16] But Buenos Aires is, in large part, a wired city. The services inhabitants and visitors can access via Internet cafes and libraries – not to mention personal computer and Wi-Fi set-ups linking users to the Internet – vastly outnumber access points now existing to the BAL network. With outdated technology and limited participation, the nodes in BAL offer comparatively few connections – and, as mentioned, there exist a large handful of nodes with no current connection to the larger network (the organizational response to questions about this is the frequently repeated maxim: "if you want the network in your neighborhood, build a node"). So why, exactly, would members *want* to build a node, to connect to the network, to participate at all if the vast majority of them have easy access to the Internet?

The answer to this question varies greatly among members: some like the fact that the personally maintained and built nodes let them avoid entanglement with corporate interests. Others say they are interested in the technology itself and claim zero interest in the political relevance of such a project. Many members with whom I spoke, however, were particularly concerned with open source and free software movements. Most shared a general belief in the freedom of information. But none of these motivations remove BAL from its strange position in the informational ecology of the city. The technology offered by the network is largely superfluous. As a content sharing system, it is limited and almost entirely duplicated on the Internet: users share music and video files that have fallen out of copyright, open-source code, images and texts that they have themselves produced. This redundancy is part of what puts BAL into the category of junk.

Thierri Bardini's fascinating *Junkware* highlights junk as conceptually liminal in cultural, political, and biological systems. He argues that junk is now *the* stuff of such systems, that is, it is the materially present facts that govern them: "Junk is the order of the day, it now affects communication, invades your (e)mailboxes, floods you with its unbearable way of just being here."[17] Junk – as we have seen in Buenos Aires – is not wholly external or excessive to urban and economic systems. In fact, it can be harnessed by these systems as the peripheral phenomena that define the center. Junk can regain value, recycled by the systems that once cast it out. Junk is amorphous. It can go the way of trash but it does not have to. It straddles the line between usefulness and the oblivion of the useless. It disturbs vertical hierarchies by refusing to be easily placed within them. BAL functions as junk partly because of the superfluity of its content, partly because it is often is built from e-waste and other trash collected from around the city, and because of the way the organization imagines itself as not yet fully useful. According to its website, BAL is built in part for a future that may or may not come into being. BAL imagines itself as a network that can be activated when natural or political catastrophe dismantles or disables other communications infrastructures – it imagines a future in which it transitions away from being junk and becomes and essential urban information system.

The BAL network, in its current state, might seem an unnecessary addition to the kind of informational exchange popularly thought to exist in a city like Buenos Aires. Because of copyright laws, users who upload content to BAL are politely reminded (though no one checks) that the content available on their nodes should be open source, creative commons, or in the public domain. This encourages the redundancy of BAL's content, weighting the network's place back toward the category of "junk." But because it is junk, it carries a trace of the systems that evaluate content, that position the network as junk to begin with. Internet users frequently come up against barriers to access when they seek protected content. Huge chunks of the global archive are either available only for purchase or via illegal access routes: through BitTorrent files, sites that operate in the hopes that they will not be found by publishing houses, etc. A BAL user, on the other hand, will notice how limited the content is on the network, but also that information is freely exchanged. No fees are required to enter, no advertisements fund sites on the network, and nearly all the material available is free of copyright. It functions as a trace by absence of the kind of commercial interests that are so ubiquitously present on the Internet.

While BAL publicly discourages the exchange of copyrighted material, no effective prohibitions actually exist. Users who put up content are responsible for their own uploads and thus each user decides the level of risk in which they are willing to engage. This has caused some debate in the organization – some would prefer an outright ban on privately owned material on the network, some believe BAL is a reasonable site for technically illegal information sharing. These questions, during the period of time I spent with the organization, were answered on a node-by-node basis. Regardless, the kind of material available and the free access granted to users highlights the strange liminality of the BAL network, its oscillation between usefulness and uselessness, that is, its junkiness.

A junk network like BAL activates its place in the city by performing its difference from more common communication and information technologies. BAL's entrance into the media ecology of Buenos Aires underscores rather than obscures certain kinds of information flows and their regulation and policing. The point is not to compete with the Internet, but to offer a different means of envisioning networks that are situated in a particular community. There is no attempt to conquer the global archive, but rather to point out that it is not a universal, all-knowledge-accessible library. The very junkiness of BAL highlights the fact that junk by the standards of global capital may indeed be repurposed, become valued once more, albeit on very different scale and for very different purposes.

I should point out too that the redundancy of content as well as the redundancy of network technology in/on the BAL network is only increasing. I was told that the organization was in something of struggle when I arrived in Buenos Aires because of political divisions within the membership. "Cataplum" lived on the outskirts of Buenos Aires in one of the *partidos*. He built his own node, which had been functioning for three months when I visited. He complained that some of the more active and vocal members of the group were trying to push it toward left-wing political activism. He was interested in the technology, he said, and believed in the software libre movements, but had no desire for the group to do outreach in the *villas*. He seemed genuinely displeased with the BAL's participation with Once

Libre, a place he saw as clearly linked to left-wing politics. Many of the members of the group, however, were particularly excited about the organization's political potential. These participants saw both the software libre and open-source movements as explicitly linked to anti-capitalist and even anarchist practices. They supported universal network access and believed that regulation and corporate licensing were anti-democratic and prohibited innovation and exploration. The tension between these two factions within the group eventually led to a splinter organization, with a handful of members who wanted to avoid any political affiliations or projects forming a new organization. They are currently in the process of building their own network, which is distinct from BAL. This division, however, is productive in terms of envisioning the relationship between horizontal action and urban sprawl. Rhizomatic engagement with the city necessarily engenders other parallel practices, it both links places and people and produces new forms of connection. A multiplicity of autonomous networks works for, rather than against, horizontality as both a material and political (re)production of the city. Even if the new network adopts a vertical structure as it operates alongside BAL, the informational landscape and proliferating forms of community formation tied up with it, resist a purely vertical view of either the city or the movement of informational flows within it.

Gilles Deleuze and Felix Guattari's *A Thousand Plateaus* offers a useful gloss in its first and second principles of the rhizome, "connection and heterogeneity: any point of a rhizome can be connected to anything other, and must be. This is very different from the tree or root, which plots a point, fixes an order."[18] The BAL network functions rhizomatically in both the sense of its technological and organizational structure (horizontal) and in the sense that it works to bring *both* connection and heterogeneity into being in the city in ways that resist a vertical ordering topology. That the organization is splintering, shifting, and proliferating only strengthens its capacity to situate itself as distinctly and always multiple.

In this regard, the BAL network can be imagined as a functional assemblage. Though the term is appropriately difficult to pin down in Deleuze and Guattari's work, they write that an assemblage can be thought of as "an increase in the dimensions of a multiplicity that necessarily changes in nature as it expands its connections."[19] While it may not be possible to claim that the work BAL does is carried out in the full spirit of radicality suggested by Deleuze and Guattari, certainly there are clear ways that the network gestures hopefully toward the kinds of alternative forms of being and knowing that are so central to the work of *A Thousand Plateaus*.

Junk, trash, and trouble in Buenos Aires

During the months I spent with the group, the organization actively encouraged the repurposing of junked materials in the building of nodes. The Once Libre workshops were particularly geared toward this end. The pirate radio station I visited planned to use an antenna they built at one of the workshops out of a refashioned metal shipping tube meant to hold a bottle of wine. Many of the group's members are always on the lookout as they wander the city for old computers and equipment, discarded by previous users and ready to be rebuilt for the BAL network. Founding member Vampi explained that much of the network could be built entirely on old, junked equipment, particularly because many of the organization's members had ample experience constructing and repairing machines.

The complex conceptual and material troubles that arise from trash – by which I mean material, inorganic things: candy wrappers and empty cans, discarded newspapers, etc. – are no less present in Buenos Aires than they are in cities everywhere. Unlike some of those cities, however, Buenos Aires has only recently begun to develop a widespread, municipally controlled recycling process. It depends instead on an informal (though government-sanctioned) repurposing waste economy. The *cartoneros* – most of them very poor, immigrant men who live in the *villas* – work their way around Buenos Aires in the early morning hours with enormous carts packed with cardboard, old toys, trashed electronics, and any potentially recyclable plastic.[20] They sell these materials for small sums to various companies and cooperatives who then process and resell their haul. (Among the products of this repurposing are, of course, the books made by the Eloisa Cartonera cooperative).

The *cartoneros* are themselves in a junk position in terms of urban infrastructure and economy. But without them, the city would have a considerably harder time dealing with the waste that piles up, daily, around Buenos Aires – indeed their trade exists because the urban system fails to wholly order urban trash. Before these men and women arrive to sift through it, there exist vast pile-ups of junk – not yet trash but neither yet repurposed – that would likely overwhelm the city's waste management systems. But *cartoneros* do the work they do because certain economic forces in and beyond Buenos Aires have excluded them as laborers – the city has effectively cast them away as junk. This strange feedback loop is typical of junk. The systems of valuation that junk a given object, creature, subject can often take that same object/subject/creature from its position as junk and revalue it, recycling for its own purposes the thing it has cast off.

Seeing junk in this way leads us into the difficult territory of labor. I have outlined the way that the work done by BAL members and users is geared toward a radical, material epistemology of Buenos Aires and of networked communication, but we can also conceive of their work as immaterial labor. Maurizio Lazzarato defines this concept "as the labor that produces the informational and cultural content of the commodity,"[21] and this aspect of production can exist outside of the frames of what is normally considered work. The manufacture of market trends via public opinion or, the production of a certain kind of aesthetic and cultural demand fall under this category. And while advertisers or affective laborers may be functioning in this realm, so too do workers outside the workspace – those who "like" commodities or companies on Facebook, leave a data-trail around the city as they carry their smartphones to cafes and send geo-tagged Tweets about where they are and what they buy. BAL may not be producing a commodity per se and the various traces it carves out in the city are less visible to companies who mine and sell such information, but it is inaccurate to situate the network as external to global capital.

While autonomous networks, particularly urban, junk networks, indicate and perform opposition to capital's appropriation of the flow of information as a means to produce surplus, BAL also fits easily within the framework of cultural labor that is appropriable by capital. There is a hipness to a counter-cultural "tech" hobby and – at least for those participants who want to build nodes out of new materials – the group can encourage clear consumption practices just as it contributes to the training of users to build networks. Users could just as easily take that training, their own technological mobility, back to their jobs as they could employ it in furthering the autonomy or scope of the BAL network. Artists, programmers, engineers, and even simply the users of the BAL network have the capacity to be productive as much *for* capital as they might be deemed to be productive against it. As Lazzarato writes, "consumption is no longer only the realization' of a product, but a real and proper social process that for the moment is defined with the term *communication*."[22]

This aspect of BAL's participation in the city makes the junk nature of its work all the more theoretically potent. Junk's liminality – mobile within (if sometimes on the outskirts of) urban and economic systems – means that it may bear with it both the means to support such systems and the instabilities that might undermine them. If some of the work BAL members do is immaterial labor, capital works to recycle that labor, wresting BAL from junk status and reinserting the organization within the global flows of the system. The network does not suggest a way to pull users or content all the way out of the system. Instead, BAL can be harnessed to build links within the system that are ordinarily discouraged, links that are indeed hidden or refused by capital. In short, networks like BAL can cause trouble by their very existence as junk.

Such possibilities for trouble-making are also the source of pleasure to some BAL participants, suggesting that trouble-making can be a form of urban play. While it often created conflict at membership meetings, several users expressed outright joy in the possibility of using the network to fight against corporate information control or to illegally trade information. Some even stated their opposition to capital as a global financial system and argued that networks like BAL could work to overthrow, or at the very least provide some alternatives within, such systems.

One aspect of BAL's junk-y troublemaking can be seen in the potential it has to engage the built environment. Rem Koolhaas' idea of *Junkspace* highlights the spectacular nature of commercial architecture (which is certainly *the* architecture of sprawl).[23] Junkspaces are built to encourage consumption, to distract with the spectacular as they guide bodies toward endless new transactions. BAL can activate the surfaces of such spaces in alternative ways. Why not sneak a node onto towers atop a shopping mall, after all? While this may not change the experience of shoppers within the bowels of the building, it certainly uses the materiality of junkspace for a very different kind of communication – one that is not transactional, not geared toward surveillance, but rather personal, local, and communal. If BAL can begin to highlight exactly how rooftops, corners, satellite towers and the like can be charged, materially, for localized knowledge and community formation as it offers a network for sharing information, perhaps junk networks can alter the space from the outside in – perhaps they can leave traces beyond the data-trail of credit charges Koolhaas sees as the only history of junkspace.

This perspective on the relationship between junk, capital, and the urban landscape of Buenos Aires draws from the work of Argentine literary and cultural critic Beatriz Sarlo, who has written extensively on the city. In her essay, "Buenos Aires, the peripheral metropolis", she aptly describes contemporary Buenos Aires as both a troubling sprawl and an imagined landscape:

> Today the city seems to have lost its qualities, not just in the dreaded anonymity of the crowd, but by virtue of the narrative hypothesis of a return to the plains from whence it arose. Several processes are apparent here. First is the exodus from the city to the suburbs of the economic elites and the sections of the middle classes who were able to adapt or prosper during the years of neoliberal transformation. Second is the conversion of the city center into tourist spaces (where large international hotels are built); areas transformed into museums (chosen for their picturesqueness and embellished after their original inhabitants are expelled); and completely run-down districts where traveling vendors proliferate – those excluded from the labour market and the homeless.[24]

The city-actual and the city-imagined are not just the sites of political and economic transformations but also serve to order people and spaces across Buenos Aires. The urban landscape (again, both real and imagined) helps to define the places and subjects that are valued and those that are junk. Sarlo's focus on how the city is imagined in relation to the plains also highlights the felt precarity of urban life – that is, the danger that the city itself or some of its neighborhoods will fall into the category of junk, of something that can be jettisoned without any sense of loss. The BAL network can be seen as a material articulation of connectivity that works against this narrative hypothesis. BAL imagines the city as the site of possibility for connection (both informational and communal) that it certainly is. Its reach toward junked populations, neighborhoods, or even computers has the potential to destabilize some of the normative valuations made by larger urban systems and dominant urban narratives.

Junkworld

There is, finally, a very particular kind of speculative act that BAL performs. An undercurrent in the work of the organization has its eye on the coming of an entirely junked planet – a catastrophic future in which the vital communication and information systems collapse, leaving us all with only the links we have or can build ourselves. On more than one occasion during my time with the members of BAL, I heard expressions of concern about the eventual failure of the global network embodied by the Internet. According to some members, when that catastrophe comes (and it is very much imagined as catastrophic, even apocalyptic), their network might survive to link its users to a certain kind of technological and social past, and to each other. This view suggests that what appeared as the unbreakable linkage between people made through contemporary technologies, in this dystopian-cum-utopian imagination of the future, will be replaced by the clunky antennas and leaning towers of the BAL network alone. The network would then offer us the chance to build our communication systems anew, this time more equitably and inclusively. This aspect of the BAL project positions it in the history of thinking on the urban. One parallel is to the description of the apocalyptic turn in thought about Mexico City described by Carlos Monsiváis in *Los rituales de caos*.[25] He sees in the real and imagined "apocalyptic condition" of the city a spectacle that can offer both pleasure and pride to its residents. BAL members who imagine an apocalyptic future Buenos Aires, one that could erupt into being at any moment, are proud of their participation in the post-catastrophe city, when infrastructure will be made by and for the people. Where vertical hierarchies empowering some at the expense of others were, horizontal networks and communities will be.

In their writing about Buenos Aires, Gorelik and Silvestri, note a much larger trajectory in the relationship between imagined, impossible futures and the urban landscape:

> There is no room for utopia in the actual construction of a city: its processes span over long periods of time and are made of small additions, of fragments of projects that have been both transformed and distorted through different urban actions, of interminable expropriations, of changes in functions and needs. Yet, the real city does not exist without Utopias, without the prospect of ideal futures. Behind Central Park lie the transcendentalist dreams of an alternative, natural way of life in the city, as utopian as the capitalist speculative processes that turned New York into the fetish of metropolitan modernity. The relationship between utopia and the city could be summed up in this apparent paradox. It is an inevitable relationship given that utopia was not only always conceived of in the city, but rather it was conceived of *as* the city, generating organic associations between form and society.[26]

BAL's apocalyptic imagination of the city gestures toward a utopia that emerges from the ashes of catastrophe and imagines itself as an actor whose material work in the present might help usher that utopian Buenos Aires into being.

This may be seen by some as a somewhat hyperbolic reading of BAL's place in Buenos Aires. Post-apocalyptic urban communication as a picture of the network's future is not universal among BAL's users and members. At the very least, though, on the BAL wiki, the network is described as a key communication medium in case of catastrophe or, a "medio de comunicación alternativo ante catástrofes (dónde los medios convencionales suelen saturarse)"/"alternative communication medium in case of catastrophe, in which conventional mediums are flooded."[27] Decay and destruction, particularly of the ecological sort, are never far from the conceptual and material spaces associated with junk. John Scanlan has written extensively on the relationship between garbage and death,[28] pointing out that the trouble with garbage is that it reveals our inability to master nature, control time, or halt decay. And given what we now understand about global connectivity in economy, communication, and culture, this association has become one of life and death on a planet-wide scale, and one that has managed to charge not just garbage, but all kinds of peripheral phenomena, that is, all kinds of junk, with incredible force.

Nigel Thrift has aptly pointed out that some of the darker thinking so often associated with both representations of and theorization about urban life might have, as the other face of its Janus head, a politics of hope.[29] Thrift's argument is quite different from my own. He would like to acknowledge the misanthropic character of urban life as a site of pleasure as well as fear. I am suggesting that catastrophic thinking, on the terribly grand scale of natural disaster and nuclear war is, in the case of BAL, a way to imagine the possibility for rebuilding urban life. If BAL imagines a catastrophic future, it does so in order to call attention to how it is positioned to rebuild in the service of a horizontal, shared communication that does not decide what information is good or bad or what connections between communities and people are appropriate. BAL does not organize either its content or its member in a hierarchy. Its view of a future in which it is the surviving communication medium is a chaotic, nearly anarchic picture of the world – instead of the neo-liberal vision of an entirely ordered, all-inclusive global economy (a vision aligned with many vertical hierarchies), BAL sees in a junked future the fits and starts, the loopholes and possibilities, that are now coming into being. If BAL members sometimes imagine infrastructural shut down, economic collapse, or environmental catastrophe, this may be understood as an attempt to imagine a space in which the network they are building can be harvested to remake our world for the better. Its sometime picture of a junked planet may be its cognitive map of the world we currently share, and a way of exploring the possibilities–already present–of bringing a different city into being.

Sprawl ends, junk beginnings

Thrift closes the seventh chapter of his *Non-Representational Theory* with a discussion of ethics. "An ontological ethics", he writes, "requires that much more attention should be given to the responsibility of cultivating intelligence and invention, broadly conceived as environments that are made up of informed materials which maximize instruction."[30] Thrift's ontological ethics are working specifically against the kind of vertical ordering of the world attempted by any number of troubling systems rooted in the Enlightenment. So too do BAL's vision of the city and its organizational and technological structure. While BAL may be an organization devoted to free and communal information sharing, it does not labor under those dangerous ideologies of global access and total order. The Buenos Aires envisioned and engaged by BAL is untidy. Its goal is to dive into this troubled and troubling world with its troubling and troubled cities, and create new and alternative ways of sharing and building knowledge, community, and information.

Nestor García Canclini, writing on the forces and effects of globalization in Latin America, also prescribes an alternative to the neo-liberal narrative of the word:

> We need cultural, political and media programmes with a different vision of social and cultural integration, which take into account the narratives of ethnic, national and regional cultures ignored by globalisation. The supranational integrations which have been negotiated and approved too hastily (NAFTA, MERCOSUR) only facilitate the free circulation of capital and goods, and do

> little or nothing to enable citizens to travel freely or to appropriate goods and messages in a more egalitarian way.[31]

It is not overreach to say that BAL serves as a prototype for the kind of program García Canclini hopes for. At a very basic level, it allows for the local and the global to interact without burying or eliding difference or harnessing it to sustain asymmetrical economic structures. It works to put the infrastructure that allows messages to move in the hands of its users in a conceptually and structurally egalitarian way.

In closing, it is also worth returning to the proposal Guattari made in "The Three Ecologies":

> Computerization in particular has unleashed the potential for new forms of 'exchange' of value, new collective negotiations, whose ultimate product will be more individual, more singular, more dissensual forms of social action. Our task – one which encompasses the whole future of research and artistic production – is not only to bring these exchanges into existence; it is to extend notions of collective interest to encompass practices which, in the short term, "profit" no one, but which are, in the long run, vehicles of processual enrichment.[32]

BAL is directly invested in using network technology to facilitate alternative forms of exchange and community. It invites a kind of engagement with the technology and the urban landscape that it sees as materially co-extant with the network itself. Its horizontal structure is built specifically to open up space for multiple articulations of both the city and the individual desires, messages, and knowledges of its users. BAL points to junk and sprawl as important sites of praxis. This is no small gesture.

Notes

1 Screenshot by the author.
2 Mike Davis, *Planet of Slums* (London and New York: Verso, 2006), 37.
3 Adrián Gorelik and Graciela Silvestri, "The Past as the Future: A Reactive Utopia in Buenos Aires," in *The Latin American Cultural Studies Reader*, ed. Ana Del Sarto et al. (Durham: Duke University Press, 2004), 429.
4 "INDEC. Censo Nacional de Población, Hogares y Viviendas 2010," Census (Buenos Aires: Instituto Nacional de Estadística y Censos, 2010), https://www.indec.gob.ar/indec/web/Nivel4-Tema-2-41-135.
5 The BAL wiki and organizational site have been dead for several years, though the organizational Facebook page suggested active meetings as recently as 2017. It is possible that the network has updated infrastructure or changed protocols, but I have not been able to track down any indications of such a shift.
6 Buenos Aires Libre, buenosaireslibre.org (Accessed September 2, 2014).
7 See Marina Sitrin, *Horizontalidad: Voces Del Poder Popular En Argentina* (Buenos Aires: Chilavert, 2005).
8 Eloisa Cartonera, eloisacartonera.com.ar (Accessed June 24, 2020).
9 buenosaireslibre.org (My translation).

10 Marc Augé, *Non-Places: An Introduction to Supermodernity*, trans. John Howe, 2nd ed. (London and New York: Verso, 2008), xiii.
11 Screenshot by the author.
12 Photo courtesy Buenos Aires Libre.
13 Screenshot by the author.
14 Paul Virilio, *The Information Bomb*, trans. Chris Turner (London and New York: Verso, 2006).
15 Keller Easterling, "An Internet of Things," *E-Flux*, no. 31 (January 2012), https://www.e-flux.com/journal/31/68189/an-internet-of-things/.
16 Rhizomatica, rhizomatica.org (Accessed June 24th, 2020).
17 Thierri Bardini, *Junkware* (Minneapolis and London: University of Minnesota Press, 2011), 9.
18 Gilles Deleuze and Felix Guattari, *A Thousand Plateaus*, trans. Brian Massumi (Minneapolis and London: University of Minnesota Press, 1987), 7.
19 Deleuze and Guattari, *A Thousand Plateaus*, 8.
20 Cristina Reynals, "De Cartoneros a Recuperadores Urbanos," *Consultaría CEDES*, 2003, http://www.lasociedadcivil.org/wp-content/uploads/2014/11/reynals.pdf.; and Débora Gorbán, "Trabajo y Cotidianeidad: El Barrio Como Espacio de Trabajo de Los Cartoneros Del Tren Blanco," *Trabajo y Sociedat* 8, no. 7 (2006): 1–23.
21 Maurizio Lazzarato, "Immaterial Labor," in *Radical Thought in Italy: A Potential Politics*, ed. Paolo Virno and Michael Hardt, trans. Paul Colilli and Ed Emory (Minneapolis and London: University of Minnesota Press, 1996), 133.
22 Lazzarato, "Immaterial Labor," 141. Original emphasis.
23 Rem Koolhaas, "Junkspace," *October* 100, Spring (2002): 175–90.
24 Beatriz Sarlo, "The Modern City: Buenos Aires, the Peripheral Metropolis," in *Through the Kaleidoscope: The Experience of Modernity in Latin America*, ed. Vivian Schelling, trans. Lorraine Leu, Critical Studies in Latin American Culture (London and New York: Verso, 2000), 122.
25 Carlos Monsiváis, *Los Rituales de Caos* (Mexico City: Ediciones Era, 1995).
26 Gorelik and Silvestri, "The Past as the Future," 428. Emphasis original.
27 Buenosaireslibre.org. My translation.
28 John Scanlan, *On Garbage* (London: Reaktion, 2005).
29 Nigel Thrift, *Non-Representational Theory: Space, Politics, Affect* (London and New York: Routledge, 2007), 198-200.
30 Thrift, *Non-Representational Theory*, 169.
31 Nestor Garcia Canclini, "Contradictory Modernities and Globalisation in Latin America," in *Through the Kaleidoscope: The Experience of Modernity in Latin America*, ed. Vivian Schelling, trans. Lorraine Leu, Critical Studies in Latin American Culture (London and New York: Verso, 2000), 50-51.
32 Felix Guattari, "The Three Ecologies," trans. Chris Turner, *New Formations* 8 (Summer 1989), 146.

Interlude 1

Urban/sprawl

Urban context and contestation

The forces against which urban interventions of many sorts are acting and out of which they are formed are difficult to clearly map, but there are a few ways of sketching them through 20th- and 21st-century thinking on media, power, and the possibilities of resistance. What feels like an awfully long time ago Critical Art Ensemble called "the political and cultural activist (anachronistically known as the artist)" to virtual arms, asking them to create "disturbances" in what they claimed was the new mode of contemporary power.[1] The power they described was without solid structure; it was "liquid," "nomadic."[2] Fredric Jameson names this global system directly. He argues that global capital is *the* system that governs the planet–one we can no longer see (if we ever could), certainly not one we can *represent*, but one we might nonetheless be able to cognitively map.[3] Gilles Deleuze's 1990 "Postscript on the Societies of Control" marked the transition away from surveillance and discipline as outlined by Michel Foucault and pointed instead to "a system of variable geometry," power as modular and mobile, networked.[4] Benjamin Bratton has recently picked up this line of thinking and expanded it to what he calls "The Stack," arguing that the world is now run under the regime of "planetary-scale computation."[5] These trajectories of thinking about power as both increasingly flexible and global have been paired with notions of resistance that have to acknowledge the passing of traditional political voices and binding agents. No longer can classes, races, genders, ethnicities, or nationalities be effectively leveraged (alone) as the appropriate contestational political bodies. Power moves in and around, gobbles up, translates, and repackages those forms of resistance. Power *includes* resistance, as it turns out. Deleuze and Critical Art Ensemble, among others, sought to bring into relief the possibility of a legitimate opening up of space for alternative radicalities and political power in a world that is forever crisscrossed with the mobile lines of power.

Though many of these theories of contemporary power and politics emerged in the early 1990s they are still present in critical thought on media and politics. This is in part because while the world since then has certainly changed, power has grown more complex, more dynamic, more mobile – its reach has extended rather than retreated just as the transportation and communication networks, the flows of labor and goods, and *life* in the biological and social senses, have proven more and more difficult to track. Let's not forget, however, that "treading water in a pool of liquid power," Critical Art Ensemble writes, "need not be an image of acquiescence or complicity."[6]

It seems as the pool has widened and deepened, and those invested in changing the shape of economic and social relationships need to pay particular attention to the ripples that disturbances can create. One way to understand the workings of that now wider and deeper pool in the urban context is to think of it as sprawl. Sprawl is the contemporary condition of urban life. This is not to suggest that all cities understand themselves as sprawling, or can be characterized that way by an urban geographer or a city planner. Rather, sprawl, as both a way of describing extension and expansion and an increasingly common material landscape, offers an apt way to understand what a city is, how a city is inhabited, and how it works (or fails to work).[7] Sprawl, too, may help us imagine what a city could be tomorrow.

It was Los Angeles County that inspired sociologist William Whyte to coin the phrase "urban sprawl" in 1958.[8] The view of the expansive development between Los Angeles and San Bernardino from above led him to elaborate the concept for *Fortune*, though "sprawl," without the adjectival qualifier, had been used to describe London's expansion in *The Times* a few years earlier.[9] And, of course, urban sprawl has been the cause of debate and discomfort in thinking on the city throughout the 20th and 21st centuries.

As I pointed out in Chapter 2, Urban historian Adrián Gorelik and architect Graciela Silvestri write that "the entire modernization of Buenos Aires could be limited to the question of its urban sprawl: how to control, characterize, and define it, although the urban sprawl itself was unintended and never part of the project."[10] This could be said of cities across the Americas as well: Los Angeles and Miami, the other sites of the central case studies in this book, are but two among many possible examples. To again quote Mike Davis, "sprawl has long ceased to be a distinctively North American phenomenon, if it ever was one."[11]

Sprawl is also an epistemological mode imperative for approaching the contemporary city. Henri Lefebvre's proposal to think about urban life in his *The Right to the City* in 1968 could be said to have called for a kind of sprawl:

> Knowledge must envisage a considerable number of methods to grasp [the urban], and cannot fasten itself onto a particular approach. Analytical configurations will follow as closely as possible the internal articulations of this 'thing' which is not a thing; they will be accompanied by reconstructions which will never be realized. Descriptions, analyses, and attempts at synthesis can never be passed off as exhaustive or definitive.[12]

What contemporary scholars writing on the city like Shannon Mattern, Ash Amin and Nigel Thrift, Lars Lerup, Keller Easterling and others have offered recently, as they engage with and partially answer Lefebvre's call, are rigorous approaches to the city that *sprawl*: across disciplines, across epochs, across geographical boundaries. The city itself sprawls, and so too, as Lefebvre insisted, must the models scholars use to tackle it.

Sprawl, both noun and verb, also has potent resonance with a number of critical tools scholars have used to think through modernity and its aftermath and the ways in which we might resist repressive structures as well – and this has been both within and beyond the city. Benjamin's constellation is a kind of sprawl. Deleuze and Guattari's rhizome has sprawling capacity. Donna Haraway's cyborg sprawls. So too does the commons as proposed by Hardt and Negri.

Sprawl has close affiliations, as well, with the more contemporary example offered by Bratton's *The Stack: On Software and Sovereignty*. Bratton aims to offer a partial map of "planetary-scale computation." He writes: "As the shape of political geography and the architecture of planetary-scale computation as a whole, The Stack is an accidental megastructure, one that we are building both deliberately and unwittingly and is in turn building us in its own image."[13] Urban sprawl is a large (if not global-scale) system composed by a multiplicity of active agents who contribute to an accidental but also active geography and, like The Stack, it is produced both deliberately and unwittingly and recursively shapes both the city and those who use it. But unlike Bratton's Stack, sprawl, I argue here, insists on a return to the local and the site-specific. The way one city sprawls will not be the way another does, even when it is enabled and expanded by the replication and reproduction of urban forms.

With its mandate for a return to the specific and the local as well as its much more conceptually and globally broad capacities, urban sprawl links to The Stack (one layer of which Bratton calls "the city layer") and also to what Keller Easterling describes as "infrastructure space." This kind of space, the primary space of urban sprawl is, Easterling says, "a medium of information. The information resides in invisible, powerful activities that determine how objects and content are organized and circulated. Infrastructure space, with the power and currency of software, is an operating system for shaping the city."[14] Repeatability, in the terms of infrastructure space, is essential to both the aesthetics and global spread of urban sprawl. (Easterling gives the example of the elevator as a multiplier – one of the active forms of infrastructure space. Its invention made possible the skyscraper, and in turn shaped the skyline of the modern city, worldwide). Where both Easterling and Bratton are seeking to understand some of the more universal urban phenomena, these larger overarching organizations remain in tension and play with on-the-ground, hyper-local phenomena, just as within the city as a structured and structuring environment there is a dialectical tension between, as Lefebvre described them, heterotopia and isotopia. Harvey's gloss on Lefebvre's discussion of the relationship between the two puts it nicely: "In *The Urban Revolution* [Lefebvre] kept the idea of heterotopia (urban practices) in tension with (rather than as an alternative to) isotopy (the accomplished and rationalized spatial order of capitalism and the state), as well as with utopia as expressive desire."[15]

De Certeau and his discussion of the distinctions between tactics and strategy similarly understand the city as enmeshed in dialectical tension between the poles of top-down ordering phenomena and bottom-up quotidian practices. Strategy, in his work, is aligned with large-scale powerful forces, and the view from above: the military, the city as a governing organization, global capital – these are the masters of strategy. Strategies design and delimit environments, they offer opportunities for movement or prohibitions against it. Tactics, on the other hand, are bottom-up. He writes: "Lacking its own place, lacking a view of the whole, limited by the blindness (which may lead to perspicacity) resulting from combat at close quarters, limited by the possibilities of the moment, a tactic is determined by the absence of power just as a strategy is organized by the postulation of power."[16] In the terms of urban sprawl, we can see how the pedestrian would be, exactly as de Certeau understood her, in the realm of the tactical. Her movements, her "spatial stories," as de Certeau put it, take advantage of opportunities afforded her by the sprawl, and work against some of the sprawl's larger systemic efforts to contain and police her. She takes a shortcut through an alley. She illicitly graffitis a private wall. She throws seed bombs onto highway medians. She picks fruit from branches hanging over the walls of a gated community. Her tactical work is local, even as it speaks to, and against, the much larger strategic operations imposed by the city as a system. It is the tactical (which pulls always in the direction of the local) that works in tension and relation with the kinds of strategic global systems of The Stack or infrastructure space. While the tactical approach we might locate in the pedestrian, the activist, the artist, or others is by no means novel (the situationists, among others, are commonly cited in urban studies and art historical scholarship as urban tacticians), the forces and counterforces of tactical intervention have new valence in the face of the megastructures now governing much of city life. The Stack, its city layer included, and the infrastructure space that is the architecture of sprawl are, if we take Easterling and Bratton seriously, very much tied to the structures of contemporary power.

The pedestrian's work can be both complicit with *and* counter to the larger structural impositions of the sprawl, and she is both shaped by and shaping that structure. Such positionality is in part why Lefebvre described urban practices and orders in dialectical terms. Eventually, in *The Production of Space*, he went further still, describing three dialectically related phenomena of space as it is practiced, as it is conceived by ordering mechanisms such as the state or capitalism, and as it is imagined, desired, or remembered by those who occupy and use it.[17] All three mutually imbricated fields have a tendency toward expansion that we could call sprawling. Imagining the city brings it both toward an inside, an intimate personal and individual internality, and an outside, in that it shapes the way any individual moves through a city, discusses it, writes about it, or dreams of it. The city practiced in turn spills into the material shapes and rhythms of the landscape. The city conceived by capitalism is meant to continue to sprawl in the development of ever-increasing surplus: vertically moving lines on profit graphs, forever.

The city as/and technology

As the work of Easterling, Bratton, and others whose scholarship is invested in the city today makes clear, there is no way to engage sprawl without thinking through its relationship to, or its function as, technology, broadly conceived. Easterling and Bratton understand the urban world as informational and position information as material. Fredrich Kittler goes further still, arguing that the city is, in essence, a sprawling media apparatus:

> What strikes the eye of the passerby as a growth or entropy is technology, that is, information. Since cities no longer lie within the panopticon of the cathedral or castle and can no longer be enclosed by walls or fortifications, a network made up of intersecting networks dissects and connects the city—in particular its fringes, peripheries, and tangents. Regardless of whether these networks transmit information (telephone, radio, television) or energy (water supply, electricity, highway), they all represent forms of information. (If only because every modern energy flow requires a parallel control network.) Even in those unthinkable times when energy still needed beasts of burden like Sinbad and information required messengers like the first marathon runner, networks existed. They just hadn't been built yet or, in technician's jargon, implemented. The narrow, rugged mule trail was replaced by the railway and the highway, which in turn have been replaced by no less transient copper and fiber optic cables.[18]

In Kittler's view, the city is and always has been a medium. Though the sprawl I discuss here is opposed to the deterministic model to which Kittler ascribed, there can be little question that information technology in its broadest (most sprawling?) sense is inseparable from the city itself as both a concept and a lively and specific landscape. When an activist, an artist, a politician, or – de Certeau's hero of the city, a pedestrian – intervenes, they intervene in the communicative capacities of the city. They are themselves part of its apparatus, even when their aim is to gum up its infrastructure or alter or undermine the message of its planners and police. All three of the projects detailed in this book are clear examples of this. YOUR literally writes on the material surface of the city. BAL builds and shapes informational infrastructure. Pigeonblog retools the informational capacities of its ecology.

One way to think about technology as it relates to sprawl is to reckon, as Scott McQuire has done, with the way the expansion of technology within urban space has recursively impacted the way we use space, think about it, and shape it. His term "geomedia" is meant to contend meaningfully with the way media and space interact in the contemporary city. "As much as digital media enable *emancipation* from place," writes McQuire, "they have also become a key modality of contemporary *placemaking*."[19] This is one way to say that technology itself sprawls, massively expanding within the city, and making the city itself expand (the emancipation from place of which McQuire speaks can be found in global telecommunication, in so-called telecommuting, in the expansion of trade, etc.) into globally connected territory. It is also to say that the local has decidedly *not* been evacuated in this sprawl.

McQuire's geomedia, to put this in clearer perspective, is characterized in part by positionality. The vast multiplication of GPS-equipped devices in the hands of individuals means that the mediation of our urban experiences is connected to a global infrastructure, and even, as Lisa Parks has observed, to an extraterrestrial space via the satellites circling the planet.[20] That vertical mediation described by Parks may involve wildly sprawling territory, but it also always returns and reverberates in the local, directing and misdirecting bodies in the landscape, tagging digital material with its hyper-specific place and time, and mapping movement of both people and things over time. Sprawl then, in the terms of geomedia, manages to be both very much global and very much local.

Sprawl in peril

Urban sprawl has long been lamented by urban researchers and civic activists, but what it is and what risks it both faces and what damages it may cause have changed. Long seen as an aesthetic blight, an alienating development phenomenon, a material consequence of the so-called American Dream, urban sprawl is now also the territory of disasters of a different order. The environmental consequences of urban sprawl have already been grave. Increased emissions with increased car use, toxic run-off, diminished wildlife habitat and more are all tied to sprawl.[21] Scientists have linked sprawl to increased temperatures in so-called urban heat islands. According to the U.S. Environmental Protection Agency, "the annual mean air temperature of a city with 1 million people or more can be 1.8–5.4°F (1–3°C) warmer than its surroundings."[22] And sprawl is recursive: the threats to the environment it causes in turn create threats to the sprawled city landscape. A Pro Publica study of Hurricane Harvey, for one particularly glaring example, cites urban sprawl as an essential precursor to the catastrophe:

> Houston's explosive growth is largely to blame. As millions have flocked to the metropolitan area in recent decades, local officials have largely snubbed stricter building regulations, allowing developers to pave over crucial acres of prairie land that once absorbed huge amounts of rainwater. That has led to an excess of floodwater during storms that chokes the city's vast bayou network, drainage systems and two huge federally owned reservoirs, endangering many nearby homes.[23]

Sprawl as a material phenomenon is increasingly understood as imperiled–not in the sense that it will vanish as a consequence of urban expansion, but rather in the sense of its coming inundation, conflagration, or abandonment. The globally circulated images of Houston's residents boating, rather than driving, across highway overpasses and city streets were stark, horrifying clarion calls. So too were the catastrophic wildfires that swallowed Southern California (one of which jumped Los Angeles' notoriously traffic-clogged 405 freeway) in 2017. Katrina, Irma, Maria, and Dorion all produced life-ending, city-decimating winds and storm surges.

Coastal cities like Buenos Aires, Miami, and Los Angeles all have to face the challenges of climate change or suffer devastation, disaster, even erasure. For these reasons it is essential, if sprawl is to function as both an epistemological strategy that recognizes its counterpart in a material urban phenomenon, to think through the recursive relationship between cities and climate.

Sprawl has this capacity, however, if we follow Lefebvre's call to untether our urban policies and practices from the more static visions of developers and corporate interests. This requires choosing creativity over creative destruction, positioning the city not as a singular system or circumscribed geographical location, but as an always becoming, processual, experiential and forever alterable socio-spatial phenomenon. It requires, in short that we give sprawling thinking the capacity to respond to the increasingly complex and interrelated risks posed by a changing climate.

Capitalism and the right to the city

However accidental urban sprawl appears, it has a close elationship to the accumulative logics of capitalism. Capitalism itself could be said to be sprawling. Its requirement for perpetual growth, its reliance on the debt form, its unintended and expansive consequences, while not wholly coterminal with urban sprawl, are certainly among its engines. David Harvey, in a text titled after Lefebvre's *The Right to the City* argues that cities themselves arose as social and geographical phenomena in relation to the workings of capitalism:

> From their inception, cities have arisen through geographical and social concentrations of a surplus product. Urbanization has always been, therefore, a class phenomenon, since surpluses are extracted from somewhere and from somebody, while the control over their disbursement typically lies in a few hands.[24]

While I argue for sprawl as the critical approach to the city, sprawling has also been a directive utilized by capitalism, both as concept and as material urban form. Harvey argues that what he calls "accumulation by dispossession" has been "at the core of urbanization under capitalism."[25] Capital's accumulative expansion in the urban landscape, he says, relies on housing speculation and the development of financial tools (such as the credit-default swaps that were in large part responsible for the 2008 global financial crisis) to manage the surplus produced by capital and necessary for its continued expansion and to produce even larger populations of surplus workers who, forced out of their homes and often their jobs, are pushed into growing slums.

That capitalism relies on sprawl does not deflate its theoretical force as an approach to the city. In fact it solidifies sprawl's critical power because it can offer a view of some of the systemic operations of capital just as it can find within the city opportunities to work against capital's logic. Harvey's rearticulation of the "right to the city" is itself essentially a more sprawling approach to politics–it counters that of the current regime in which power over the city is far too concentrated: "Since the urban process is a major channel of surplus use, establishing democratic management over its urban development constitutes the right to the city."[26] He further insists that such an urban politics would activate the social sprawl of the city:

> The right to the city is far more than the individual liberty to access urban resources: it is a right to change ourselves by changing the city. It is, moreover, a common rather than an individual right since this transformation inevitably depends upon the exercise of a collective power to reshape the processes of urbanization. The freedom to make and remake our cities and ourselves is, I want to argue, one of the most precious yet most neglected of our human rights.[27]

Sprawl, when it is framed as and leveraged for a radical politics is strongly aligned with the common. As a material phenomenon it also encompasses wildly diverse neighborhoods, populations, architectures, and ways of life, which bolsters such an alignment. To think the city as sprawl, and to call for its politics to sprawl, means opening toward the differences, frictions, and diversity that make up urban life. Of course, it is this difference, friction, and diversity that make urban life compelling in the first place.

Finally, sprawl as both noun and verb, is better suited than other ways of modeling the urban to contend with the realities of the contemporary city. As McQuire writes:

> If urban space has historically been defined by the relation between static structures and mobile subjects, this dichotomy is fast giving way to hybrid spatialities characterized by dynamic flows which not only dissolve the fixity of traditional modes of spatial enclosure, but problematize the unified presence of the subject traversing their contours.[28]

Part of the work of problematizing that unified presence while still reading its remaining legitimate force in the making and shaping of the city, is facing urban sprawl, working toward a common sprawling politics, and exploring both sprawl and the works of those who intervene within it, sprawlingly.

Urban sprawl is messy. It's porous. It's abstract. It's material. And, essentially, for Lefebvre, writing when the city must have seemed filled with revolutionary potential, the city is *alterable*. More so: the city *must* and *will* be altered: "It is impossible to envisage the reconstitution of the old city, only the constitution of a new one on new foundations, on another scale and in other conditions, in another society."[29]

Notes

1 Critical Art Ensemble, *The Electronic Disturbance* (New York: Autonomedia, 1994), 12.
2 Critical Art Ensemble, *The Electronic Disturbance,* 12–14.
3 Fredric Jameson, *Postmodernism, or, The Cultural Logic of Late Capitalism* (Durham: Duke University Press, 1991), 51.
4 Gilles Deleuze, "Postscript on the Societies of Control," *October,* Vol. 59. (Winter, 1992), 4.

5 Benjamin Bratton, *The Stack: On Software and Sovereignty* (Cambridge: The MIT Press, 2015), xviii.
6 Critical Art Ensemble, *The Electronic Disturbance*, 12.
7 See the compelling study and accompanying data from *The Atlas of Urban Expansion* out of New York University: http://atlasofurbanexpansion.org/about
8 William H. Whyte Jr., "Urban Sprawl," in *The Exploding Metropolis*, ed. William H. Whyte Jr., (Berkely: University of California Press, 1993), 133–36.
9 "sprawl, n.". OED Online. June 2020. Oxford University Press. https://oed.com/view/Entry/187617?rskey=B2AOgF&result=1 (accessed June 29, 2020).
10 Adrián Gorelik and Graciela Silvestri, "The Past as the Future: A Reactive Utopia in Buenos Aires," in *The Latin American Cultural Studies Reader*, ed. Ana Del Sarto et al. (Durham: Duke University Press, 2004), 429.
11 Mike Davis, *Planet of Slums* (London and New York: Verso, 2006), 37.
12 Henri Lefebvre, "The Right to the City," in *Writings on Cities*, trans. Eleonore Kofman and Elizabeth Lebas (Oxford: Blackwell Publishers, 1996), 153.
13 Bratton, *The Stack*, 5.
14 Keller Easterling, *Extrastatecraft: The Power of Infrastructure Space* (New York: Verso, 2014), 6.
15 David Harvey, *Rebel Cities: From the Right to the City to the Urban Revolution* (London: Verso, 2012), xvii.
16 Michel de Certeau, *The Practice of Everyday Life*, trans. Steven Rendall (Berkeley and Los Angeles: University of California Press, 1984), 38.
17 Henri Lefebvre, *The Production of Space*, trans. Donald Nicholson-Smith (Oxford: Blackwell Publishers, 2000).
18 Matthew Griffin and Friedrich A. Kittler, "The City Is a Medium," *New Literary History* 27, no. 4 (November 1, 1996): 717–29.
19 Scott McQuire, *Geomedia: Networked Cities and the Future of Public Space* (London and New York: Polity, 2016), 6.
20 See Lisa Parks, *Rethinking Mediation: Vertical Mediation and the War on Terror* (New York: Routledge, 2018).
21 For a long-form read of sprawl's relationship to the environment, and to the U.S. economic model specifically, see George A. Gonzales, *Urban Sprawl, Global Warming, and the Empire of Capital* (Albany: State University of New York Press, 2009).
22 OAR US EPA, "Heat Island Effect," Collections and Lists, US EPA, February 28, 2014, https://www.epa.gov/heat-islands.
23 "When Climate Change Meets Sprawl: Why Houston's 'Once-In-A-Lifetime' Floods Keep Happening," ProPublica, December 17, 2019, https://projects.propublica.org/houston-cypress/.
24 David Harvey, "The Right to the City," *New Left Review*, no. 53 (October 2008): 23–40, 24.
25 Harvey, "The Right to the City," 34.
26 Harvey, "The Right to the City," 37.
27 Harvey, "The Right to the City," 23.
28 Scott McQuire, *The Media City: Media, Architecture, and Urban Space* (London: Sage Publications, 2008), 132.
29 Lefebvre, *The Production of Space*, 148.

3 YOUR MILLION DOLLAR HOUSES WILL SOON BE UNDERWATER

Coming catastrophe in Miami and the plastic city

"YOUR MILLION DOLLAR HOUSES WILL SOON BE UNDERWATER." In August of 2016, an ambitious, anonymous artist (or artists) wrote this declaration, in enormous, black, block letters, atop an abandoned hospital in South Miami Beach (Figure 3.1).[1] The proclamation, and the medium of its production exemplify what is at stake in imagining urban catastrophe in relation to and within vertical terrain. The work is an ephemeral piece of illicit public art and an activist response to the increasing threats posed by anthropogenic climate change to the Miami area, and to the relationship of those threats to the history of urban development in a city that favors financial flows and surplus management over the material realities and lived experiences of the urban world.

Miami is an exceptional site through which to consider the imagination of catastrophe because of the strong links in the city between urban development practices and speculation (both economic and imaginative), and the disastrous possible futures for a city already experiencing the consequences of rising temperatures and seas. The city's disavowal of climate change, so clearly embedded in its practices and political discourse (particularly under the administration of Florida Governor Rick Scott, which banned the phrases "climate change" and "global warming" in official communications from the state's Department of Environmental Protection[2]), is a position that the graffiti writers' work (hereafter referred to as YOUR) screamingly refuses.

Since Spanish colonists first sought to tame and dominate the land on which it now sits, Miami has been something of a fantastic city, developed despite, rather than because of, its geographical, social, and geological facts on the ground. T.D. Allman, in his excellent history of Florida, writes:

> The most important revelation Florida forces on us, if we are willing to adjust our perspectives, is how irrelevant our narratives are. What we consider the aberrations, the disasters, the exceptions —they are the real narrative. We are the aberrations, the ones who imagine that the hurricanes, the epidemics, the killer frosts, and the human catastrophes constantly sweeping over Florida are not the natural state of affairs.[3]

Miami, as both the exemplar and outlier city in the state, gives us a way of conceiving catastrophe as the city's underlying ethos, even if it is not directly visible

Figure 3.1 The YOUR graffiti as seen from a nearby condo tower
Source: Stephen Cohen.

in urban forms and discourse. Approaches to catastrophe range from the vertical speculation by developers whose condo towers reach increasingly upward and whose profits are expected to forever rise, to this singular piece of graffiti whose visibility depends on the verticality offered by the height of the empty hospital building. The two situate the city in contrasting lights. On one side is a disavowal of catastrophe in the form of endless profit accumulation and urban growth, on the other a reorienting of the city toward both the already catastrophic impacts of urban development under capitalism and anthropogenic climate change. Both the capitalist speculations on the financial future of the city and YOUR's speculation on the consequences of a nearing climate catastrophe operate in and engage in the vertical field. This chapter will argue that YOUR reshapes the city as plastic rather than resilient, by operating against the neoliberal logics that support notions of resilience so often touted by city developers and boosters. YOUR sees the city as plastic: that is, as capable of giving and receiving form rather than as flexible in either the material or metaphoric sense. (I further develop plasticity below and in the interlude following this chapter). The idea of plasticity understands the city's essential quality as its transformative capacity. Where resilience discourse seeks to maintain vertical power by evacuating and suppressing structural critique and inculcating subjects as flexible but not agentive, plasticity works against vertical hegemony by insisting on the processual, potential, and always active alterability of the urban world. Where neoliberal urban policies and development practices have consistently erased the catastrophic impacts of capitalism on urban life, a reframing of the city as plastic, rather than resilient opens up the urban field for resistance and leads it toward more equitable and sustainable new shapes.

Miami: Hemispheric city, informational city, transient city

Ashley Dawson's study of urban risk in the age of climate change, *Extreme Cities* takes Miami as an exemplary case: "Of all the world cities threatened by climate change," he writes, "Miami is faced with particularly daunting challenges."[4] As Dawson and others have noted, the entire region is built primarily on porous limestone, meaning the immobile borders, whether coastal or subterranean, drawn by most map-makers are largely imaginary. The fiction of solid ground is not just cartographic. Miami Beach, for example, is manmade, and most of the city was built by filling the marshes of the Everglades. Miami, in this way, underlines that cities are as much discursively produced as they are materially manufactured and that imaginative work does, indeed, help to shape them.

Miami is also an exemplary city in the age of climate catastrophe. The risks the region faces are increasingly drastic: "With just 3 feet of sea level rise, more than a third of Southern Florida will disappear below the waves. Once sea levels rise 6 feet, more than half the region will be gone."[5] Flooding isn't Miami's only possible perilous future, either. In November of 2018, the United States released its most comprehensive report on the impact of climate change to date. It laid out the consequences of a not-so-slowly warming planet in clear–and dire–language, describing South Florida as among the most endangered communities in the United States, and among those that are already suffering from the impacts of climate change. Greater Miami-Dade county faces imminent threats including rising waters, extreme heat, and vector-borne disease.[6] And, of course, with these risks come shifts in urban development and population, particularly in the form of climate gentrification.

The consequences of any number of the threats outlined by scholars, governmental organizations, and artists like the author of YOUR are, unsurprisingly, asymmetrically distributed. The urban poor (in Miami, as in most U.S. cities, this population is disproportionately composed of people of color) will continue to be the most at risk. But, as anyone driving into Miami Beach along A1A during the very brief stint in which the graffiti was up might have considered, the consequences coming will not remain entirely within the capsular borders dividing Miami's more affluent communities from everyone else.

Miami, like the other cities through which this book moves, is both a port city and an immigrant city. Its particularities, however, are essential to its position as an urban site, chief among them, the incredible demographic shifts that have taken place across South Florida in the last seventy years. These shifts, launched in large part by immigration from Cuba in the decades following the Cuban Revolution, swelled by waves of immigration from Haiti following the convulsions of the François "Papa Doc" Duvalier regime and the collapse of the Jean-Claude "Baby Doc" dictatorship, and from Latin America throughout the latter half of the century, transformed Miami from a primarily white, Southern city into one in which the city's "defining characteristic," according to geographer Jan Nijman, is transience.[7] People from across the globe move in and out of Miami-Dade county more or less constantly. Tourists arrive and depart year-round. These fluctuating populations give the city a compelling and chaotic diversity (even as that same transience also poses problems for coherent political coalition building). Miami's transience also contributes to its position, according to Nijman, as the first hemispheric city, and an exemplar of what urban landscapes across the Americas are likely to become.[8]

Its hemispheric position is exemplary not only in relation to flows of people, but also to flows of money and of information. "Miami serves as the premier router, so to speak, between the corporate capitals of the world and Latin America," Nijman writes. "There is no other region where one single city so monopolizes the transaction of flows between that region and the rest of the world."[9] The idea of the "router" here is more than just a metaphorical device. The city houses communication and computation infrastructure essential to linking the United States, the Caribbean, and South and Central America to nearly 150 countries around the world. Downtown Miami is home to the Network Access Point (NAP) to the Americas, operated by the multinational corporation Equinix. In a windowless facility, just blocks from one of Miami's poorest neighborhoods, informational traffic and computational power quietly hums away through fiber optic cables and within server cages, literally linking the city to the rest of the hemisphere, and the world. This emphasizes both that Miami, like all cities, is mediated and mediates in both literal and conceptual terms. It transmits information as a hemispheric hub and is itself an informational landscape produced by both discursive and technical forces. But it also positions the city as deeply embedded in ongoing contestation and engagement within the vertical field, as outlined by Lisa Parks. The NAP means that Miami is directly involved in the "use of aero-orbital technologies (satellites, aircraft, transmitters) and spaces (orbit, air, spectrum) to support such activities as the international distribution of audiovisual signals, the patrolling of movements on and beneath the earth's surface, and the physical destruction and reconstruction of lifeworlds from above"[10] (more on this later).

Miami serves as both a hinge site and conduit between North and South America, between Los Angeles and Buenos Aires. The landscape into which the graffiti writer intervened in 2016 is starkly shaped by the flows of people, goods, information, water, and, of course, money that circulates through the city from near and far. I see the YOUR intervention as both contending with these flows and speculating about their futures. My reading attempts to take this piece of graffiti seriously as a piece of ephemeral art, both macabre and playful, that exposes some of the broader implications of Miami as a city in relation to catastrophe and verticality. YOUR not only positions the urban environment as having a concrete relationship to a coming catastrophe but also gestures toward ongoing catastrophes already unfolding in and around Miami, primarily those of climate change and urbanization under capitalism. In Miami and elsewhere, of course, these two processes are intertwined in the form of climate gentrification, which is quite literally a racialized process of verticalization – in which primarily white and wealthy people move upward, onto higher ground, displacing communities of color as they do. In the face of this reality, YOUR carries an implicit fantasy: that somehow, despite the incredible power of capitalist urbanization, climate change will level the playing field. YOUR seems to believe or hope that the rich will not escape the consequences of unsustainable development in the face of the coming catastrophe. Part of the imaginative and speculative work done by "YOUR" is this fantastic projection.

Climate projections

Rising sea levels and ongoing housing crises are likely the most pressing challenges for Miami-Dade County in the coming decades. YOUR is clearly an articulation enmeshed deeply with these concerns, and its speculation on the outcome of both is at once grim and fantastic. To begin with this intervention's take on climate, I would like to think in terms of the piece's temporal framing: "soon," the work projects, the floods will come for the rich.

The author of the statement positions herself toward a future about which she can make knowledge claims. Climate change is often framed this way: always on the horizon but not quite in the present. This has consequences specific to Miami Beach because the rising seas aren't coming "someday": they are very obviously already here. Miami Beach floods frequently. Each king tide wreaks havoc on the city's infrastructure, clogging drainage systems, and roadways alike. Million dollar houses already have been flooded, if not wholly subsumed by the rising waters of the Atlantic.[11]

But climate consistently poses problems for concrete temporal framing. As Eva Horn has written, the epistemological challenge that climate poses is exacerbated by the present cultural milieu. Distinct from weather, Horn explains, which we can conceive, picture, and immediately experience, "climate does not take the form of an event."[12] For this reason, climate change "has come to epitomize a new type of threat that [she calls] 'catastrophe without event.'"[13]:

> Even though it can only be recorded *on site*, "climate" has by now become a *global* concept, with a focus on its long term *temporality*. Once a phenomenon of overwhelming phenomenal intensity (traveling was seen as a "change of climate"), the modern systemic concept of climate has withdrawn into the realm of abstraction and computation.[14]

This means that climate has moved into the realm of speculation. "It is no longer just a matter of statistics (as was still the case in 19th-century meteorology) but rather one of gauging *potential* system behavior over a long period of time."[15] In the case of climate science, specifically in its reliance on simulation, speculation is leveraged as a mechanism to order the future, to make it quantifiable. A "'conjectural science' that uses mathematical models in the hope of rendering the fluidity and hypercomplexity of atmospheric phenomena calculable," simulation in climate science, like YOUR's sprawling message in Miami Beach, calls out to a future it makes claims to know.[16] Both climate models and YOUR are oriented speculatively in that they imagine a particular future from within a field of chaotic uncertainty. But these two forms of envisioning the consequences of climate change leverage speculation differently.

Climate science, however deeply useful its models may be, is still processed in larger discourses as firmative, rather than affirmative.[17] It forecloses and contains possible futures by rendering them mappable, by producing chaotic and unpredictable phenomena as knowable and foreseeable. This is partly what Horn points to when she notes that climate models seek to render hypercomplex and dynamic systems legible. YOUR is equally speculative. However, its medium, its performance as graffiti, and its fantasies about the impact of climate change on urban economies, mean that its speculation remains affirmative: even as it makes a teleological statement, its context and manner of coming into being destabilize its own doomsday proclamation, leaving room for other possible futures and other possible urban forms. In some ways the graffiti piece, indeed, all successful graffiti pieces, mimics and performs the eventlessness of climate change. If they are successful, the moment of their production is erased. They don't happen, they simply are. In its graffiti form, YOUR's speculation models the eventlessness of the climate change at which it aims its thinking and gives us the opportunity to access the transformative possibilities still available in our responses to it.

Some might wonder why one would favor the affirmative speculation of YOUR over the more rigid and predictive vision offered by climate modeling. The answer is less about the value of the latter, and more about the kinds of possibilities afforded by the former. Climate change has likely material consequences that we should be aware of. But it also operates on social scales and within affective registers about which climate models tell us nothing. Among these are, as Rob Nixon has written, the slow-violence of climate catastrophe on underserved communities and underdeveloped nations. Or, as my colleague Lindsay Thomas has described, the "experience of uneven and inaccessible time, of temporalities that occur beyond the thresholds of our perception, of changes we can't see or directly experience but that nevertheless directly affect us" are very much among the impacts of climate change.[18] While climate models speculate on a future they claim to read, creative interventions such as YOUR can move us toward an understanding of climate change that goes beyond weather patterns or predicted flood lines and toward other registers, other experiences. These are affirmative speculations in the sense described by an uncertain commons: "Affirmative speculation affords modes of living that creatively engage uncertainty. Its stakes are resolutely collective: often sabotaging individuated and privatized prescriptions, it builds on the tentative mutualities that arise in the face of uncertainties. In short, affirmative speculation embraces ways of living in common."[19]

Such interventions allow us access to thicker, broader views of the nature of climatic phenomenon's effects on the actual experience of everyday urban life. They work against purely a top-down, God's-eye view of climate change and toward one that works in multiple directions and across multiple scales: economic, physical, affective. YOUR asks viewers to think about the economic impacts of climate change as well as impacts to the built environment. Because its production was illegal, and thus inevitably ephemeral, it performs the difficulty of concretely representing the city's future and highlights the ways in which some responses to climate change are legitimized within the city's juridical and governmental frameworks (climate gentrification, for example) while others are not (direct address and aid to people and groups who are not property owners).

Running parallel to the firmative speculative logic of discourse around climate science are the kinds of risk management strategies deployed by governmental and corporate interests in the city. As an exemplar, we can look again toward the Miami NAP. In its promotional materials, Equinix points out that the NAP is housed in a massive structure designed to withstand Category 5 hurricane winds. Such winds, of course, are projected to come sooner or later. The facility is heavily secured against human threats as well. According to a BBC report, the NAP "features its own X-ray machines and sniffer dogs as well as Internet-based firewalls, intrusion detectors and other protection devices. [...] As most of their information is so sensitive, no cameras or other electronic devices are allowed. Access to the third floor – 125,000 square feet entirely dedicated to US government users – is restricted to US citizens and requires government clearance."[20] Even the satellites on the roof are obscured by large domes so that their directionality cannot be gauged. These measures suggest an orientation toward a future threat on its way: climatological, social, or both, but they also offer the kinds of speculative structures that are so common to capitalist biopolitics. Strategies employed by Equinix, and by others whose aim is the continued production and management of surplus, are that of containment and capsularization. In the case of the NAP, even the architecture of the building–securitized, windowless, guarded–turns away from the city. As a manifestation of informational capitalism, it testifies in both material form and discursive production to an urban catastrophe, sees that catastrophe always on the horizon or outside the gates, but perpetually defers its arrival and promises protection from it.

Capitalism situates its own structure as the gatekeeper, keeping catastrophe at bay. This is truer in the Miami metropolitan area than perhaps in any other American city precisely because Miami is *already* sinking, already experiencing the consequences of climate change. Yet, the more present the ongoing catastrophe, the more fervently capitalist enterprises must insist that it is somewhere, and always, in the future, and never on the ground.

With its obscured satellite dishes and links to the U.S. security apparatus, the NAP also shows us the ways in which vertical hegemony, described by Parks, is also linked to certain forms of producing discourse about the city and its relationship to climate catastrophe. The same site used by the United States to conduct operations in the vertical field is also the site that promises to protect those operations from catastrophic failure in the face of a climate event, or an intrusion by unwanted upheaval of another sort within the city. The entwining of the state and capitalism embodied by the NAP flatten the distinction between human-scale intervention and climate threats. Both are framed as perpetual risks, and both are eternally deferred and denied. That the NAP is a physical site in the city that actively performs its securitization also gives material evidence to the claim that vertical power always returns to the lifeworlds on the ground – even while the discourse around it denies strong material ties to place and space in specific urban contexts.

Miami's real estate market also offers a take on verticality in the city that is exemplary in its deferrals and denials, of both the climate catastrophe the city faces and the ongoing catastrophe of a housing shortage which has, at least on Miami Beach, all but eliminated affordable housing. According to one report, the housing crisis rivals climate change as a threat to the region.[21] Dawson writes that "real estate development [is] a key motor of contemporary capital accumulation [that] generates forms of irrational and unsustainable urbanization."[22] In Miami, this is visible everywhere. A Marketplace report aptly described condos in the Miami area not as living spaces, but as "apartment shaped financial instruments": "Downtown Miami's condo explosion turns out to run on its own financial logic – far removed from rising seas."[23] The report points out that many investors in these condos, which are often still in the planning phase but not yet in construction, are foreign. They speculate on rising prices (forever on a vertical trajectory), rather than rising water.

The investors who have built downtown Miami, Brickell, and much of the Miami Beaches may or may not envision a catastrophe on its way. But operating under a free market ideology, they have little incentive to care as long as prices keep rising and they can sell before whatever looming catastrophe bursts the economic bubble. Continued investment in these and similar developments in coastal cities across the Americas require a disavowal of the current climate risks. Regardless of how supposedly resilient or sustainable a design or a building is, if even some of the most conservative models are right, much of Miami will simply be underwater. Such investment and development practices are, of course, exactly what Dawson points to: irrational and unsustainable by the terms of the vast majority of the city's population. They are rational practices, though, when seen through the lens of real estate speculators and developers in the brutal calculations that buoy certain lives and futures, very much at the expense of others. Furthermore, financial and real estate speculation as it works in the case of the Miami market is almost entirely abstracted from the worlds it impacts: that is to say, instead of material spaces and interactions in the city, it relies on what it claims are quantifiable financial futures, whose unfolding it only vaguely ties to the living city, its fabric or its communities, as those empty condos built for trading attest. In this way, catastrophe in capitalist urban development is forestalled only for some through a rationality that categorically denies the value of certain lives, certain spaces, and certain ways of inhabiting the city. Catastrophe is coming, it says, but it will be managed by the free market, for the free market, at any cost. The one system that cannot be catastrophically altered is, perpetually, capitalism itself.

Resilience and flexibility

Capitalism resists its own dismantling by insisting that not even catastrophe can undermine it. But the market inevitably does crash (again and again), and the floods keep coming. When they do, the abstraction of the city created by markets and developers comes down from its vertical heights in the form of distinctly material blows delivered to lives and spaces on the ground.[24] The double meaning of "under water" for YOUR points to both possible, and possibly inevitable outcomes: homes under water can mean both homes flooded, and homes owned by those who can no longer make payments on mortgage debt. Investors with financial interests in the condo market, though, will not feel the impacts the way that residents, especially poor residents who rent rather than own (and certainly rather than trade) will. These are the residents who will be displaced, dispossessed, and discarded. The example of New Orleans after Katrina is illustrative: an estimated 1800 people, the majority of whom were both poor and Black, died during the storm and its aftermath and, at least as of 2010, there were 118,000 *fewer* Black residents in the city than there had been before the levees broke.[25] It was then and remains these residents who are most consistently asked by municipal, state, and federal institutions, and corporations to be "resilient."

Jeremy Walker and Melinda Cooper have described the discourse of "resilience" and its close marriage with neoliberal biopolitics.[26] Cooper and Walker track the resurgence of "resilience" as a concept in U.S. neoliberal economic policy post-2007 and in the "preparedness discourse" of U.S. Homeland Security. In both, they find resilience as a state into which the neoliberal subject must situate herself. The position of the subject and the community at risk, within the context of a predicted but ultimately unforeseeable future, "a 'culture' of resilience [turns] crisis response into a strategy of permanent, open-ended responsiveness, integrating emergency preparedness into the infrastructure of everyday life and the psychology of citizens."[27] The assumption that informs resilience discourse, as Walker and Copper note, is that the free market, which itself puts subjects at risk, demands resilience but must always remain unquestioned as the governing system.

The Miami metropolitan region is an active and vocal participant in the perpetuation of "resilience" discourse. Miami-Dade county, in partnership with Miami Beach, and the city of Miami were selected as part of the Rockefeller Foundation's "100 Resilient Cities" initiative. "100RC," as it calls itself, provides funding and other assistance to its selected cities to "help [them] change the way they understand their risks and plan for their futures."[28] In their promotional materials, the foundation explains exactly what it means by "urban resilience": "the capacity of individuals, communities, institutions, businesses, and systems within a city to survive, adapt, and grow no matter what kinds of chronic stresses and acute shocks they experience."[29] While there may be many benefits to Miami's participation in the 100RC initiatives, not the least of which is a consistent acknowledgment of climate change (historically, not so easy to find in Florida), this discourse brings its own catastrophic urban imagination. If individuals, communities, and even systems are expected to display resilience, to survive and grow, "no matter what kinds of chronic stresses and acute shocks they experience," then where is there room to acknowledge the *failure* to thrive caused by structural, slow and quick violence, both racial and economic? Are those whose houses go underwater, in either the sense of financial foreclosure or flooding, essentially guilty of failing to be resilient? Resilience, it is clear, is viewed not only as a necessary quality in the contemporary Americas; it is also positioned as a moral good. The language of "no matter what" is a language that deflates possibilities of revolt and resistance. Are there systems, conditions, and outcomes that people and communities should not be asked to survive in the first place? Are there "chronic stresses" we might refuse entirely? Resilience discourse embodied by the likes of the 100RC initiative also completely refuses the possibility that some systems, perhaps, should NOT survive, or grow. Key among them in the Miami context: capitalist urbanization and climate gentrification. Growth, in the case of these systems, has been catastrophic already.

There is too, in resilience discourse, the spectral presence of repugnance. Even those who exhibit the qualities of "resilience" remain excluded from the rewards of such supposedly "good" behavior. As Thomas pointed out to me, "repugnance" is, in fact, an obsolete synonym for "resilience."[30] Her forthcoming book, *Training for Catastrophe: Fictions of National Security After 9/11*, unpacks the ways in which U.S. Homeland Security "preparedness" discourse is discomfited by resilience not because of its uncomfortably close relationship with death (though it does have that relationship) but rather because it forces preparedness to see its own power's biopolitical foundation: the homeland is constructed around the notion that those outside of it should not, in fact, be safe.[31] Urban resilience discourse operates somewhat differently than resilience within the state security context. Because it operates within specific cities, it tends not to frame resilience in the context of national citizenship or homeland. Nonetheless, it heavily leverages notions of belonging that run counter to the actual make-up of the city, refusing to acknowledge some of the very populations into which it hopes to inculcate resilience.

The 100RC initiative produced "Resilient305," a 154-page report outlining resilience strategies for Greater Miami and the Beaches that never uses the the term "Black" nor even "African American"; indeed, it never refers to "race" at all, insisting instead on "diversity." The document acknowledges "an overwhelmed criminal justice system" and an "average household income 19% lower than the national average" but refuses to do any real demographic math: no links are made, for example, between race and the carceral state, or between poverty and ethnicity, yet the Resilient305 policies claim to address "shocks and stresses in a holistic manner."[32] This utterly colorblind approach is astounding given that Miami-Dade is home to the fifth largest Black population in the United States and that that same population is wildly overrepresented in Florida's jails and prisons. Clearly resilience in this context stands on the erasure of certain communities and experiences: expressly those of Black and brown people and of the poor. Used this way, urban resilience offers a smooth, flat vision of the landscape at the expense of critical approaches. It actively avoids addressing the systematic oppression that literally imperils the lives and well-beings of many. In this way it ignores the very barriers to resilience the city's actual inhabitants face. It bypasses ongoing catastrophe in favor of perpetual flexibility for individuals, communities, and cities. The meaning of "resilience" in a materials sense is, in fact, nearly identical with flexibility: "Elasticity; the power of resuming an original shape or position after compression, bending, etc."[33] In other words, resilience's very real meaning is a return to the status quo: to go back to "normal," even if that normal is utterly unsustainable, or even deadly.

Unlike resilience discourse, YOUR engages the chronic stress of structural violence, of the ways that the city values some and not others. Take first the site of its appearance. The building, which once housed the South Shore hospital, had been empty for years at the time YOUR appeared. It and the surrounding lots are the gateway to Miami Beach: the first thing that masses of visitors, both tourists and locals, see as they drive onto Alton road to enter the city. When the graffiti appeared, the site was the subject of an ongoing struggle about urban development on the island. The conflict, which, according to *The Miami Herald* included "nearly 100 meetings and a public battle between neighborhood groups" was primarily over both the height of a planned development and the demand by some residents for more green space in the city.[34] The solution, apparently, arrived in August of 2018. What remained of the hospital was demolished as soon as permits were issued and, in that structure's place, a 44-story luxury condo building will rise[35] (Figure 3.2).[36] The city is also slotted to get a 3-acre park as part of the complex. The building on which the dark proclamation of YOUR appeared is slated to become the very homes YOUR claims are in immediate danger.

The planned housing is also part and parcel of a long-standing development strategy in Miami Beach and throughout Miami-Dade county that worsens the housing crisis by embodying the structural violence David Harvey calls "accumulation by dispossession."[37] This process, through which low-income residents and communities are perpetually displaced by gentrification, "development," infrastructure projects, and other processes is, according to Harvey, "at the core of urbanization under capitalism."[38] While the site of the YOUR intervention was not occupied by low-income housing, the entirety of Miami Beach was once

Figure 3.2 What remained of the South Shore Hospital building was demolished to make way for a new condo development

Source: Video still, courtesy The Miami Herald. https://www.miamiherald.com/news/local/community/miami-dade/miami-beach/article229323424.html

home to working class communities who were gradually driven out by the rising condos, restored Art-Deco hotels, and the soaring rents that accompanied them. At the time of this writing, the median monthly price for a one-bedroom apartment on the island is $1580.00.[39] The particular chronic stress of housing costs disproportionately affects the entire metropolitan region. A study conducted by The South Florida Housing Consortium found that in 2015 nearly 50% of households in the county were cost-burdened, meaning their housing costs made up more than a third of their income, making Miami-Dade the fifth most unaffordable housing market in the United States.[40] YOUR, by intervening precisely on the space where something *other* than million dollar homes could be built, and by pointing to uneven economic development and unevenly distributed economic gains, looks economic violence in the face in ways that resilience discourse and its models of the city and individuals within it cannot.

While affordable housing is in dire short supply, those million dollar homes to which YOUR refers are everywhere. The city is exemplary in its position as a hub of real estate speculation and consumption. As Alejandro Portes and Ariel C. Armony have written,

> In the logic of property capital, [Miami Beach's] role is exclusively one of consumption. Developers and hoteliers sell the beach by the square foot to the well-heeled buyers and tourists from everywhere and then sell this same population to store, restaurant and bar owners catering to its every whim. [...] the poor and the working class do not live in the Beach, being confined to visiting publicly accessible parks. There is no area of concentrated poverty anywhere in the city.[41]

In scrawling its foreboding message at the top of what remained of the hospital, YOUR appears highly aware of the audience to whom it speaks, perhaps, even, given that the site was a hospital, to the very real mortal consequences of both capitalist urban development and anthropogenic climate change. It seems equally aware of the relationship, underscored by many authors referenced in this chapter, between real estate speculation and unsustainable development so very entrenched in the specific landscape of Miami Beach. Resilience discourse, in contrast, ignores and disavows the relationship between these two forms of intertwined violence, instead conflating the city itself with the individuals within it, who must be forever flexible in the face of shifting economic flows or increasing climate risks. They must simply "adapt" to the fact that there may be nowhere for them to affordably reside. This is abundantly clear in the Resilient305 report, which offers only additional capitalist development to address the housing shortage and economic disenfranchisement.: "Sixty-seven federally designated Opportunity Zones have been mapped in Miami-Dade County to revitalize economically distressed communities using private investments," it claims. Offering an "Atlas of Opportunity" the document (vaguely) explains that "to yield maximum financial benefits, the private sector will be looking to make quick decisions on where to invest. The Opportunity Zone program presents an opening for community leaders to focus on resilient strategies and work with Opportunity Funds to achieve and accelerate broad-based outcomes."[42] Note that it is community leaders who are asked to focus on "resilient strategies" in order to access to private funds, and so that their communities can welcome private sector investment. City governments, in this model, do little more than advertise to investors on behalf of the disenfranchised people and places under their governance.

These communities, in Miami as elsewhere, are very often communities of color. They need more, of course, than financial investment from the private sector. In Miami and across the United States they have been systematically excluded from the benefits of so-called urban development. Indeed, that very development has often been engineered to disenfranchise and dismantle them, as was the case in the late 1960s when the majority Black Miami neighborhood of Overtown, once known as "the Harlem of the South" was bisected by interstate 95, displacing residents and devastating a once thriving local economy.[43] It is impossible in the context of U.S. cities to understand the real stakes of urban development without taking into account its relationship with white supremacy. Red lining, development and infrastructure projects like that of interstate 95, banking practices, and the gentrification trends that perpetuate the displacement and dispossession of Black communities mean that cities are not now nor ever have been environments conducive to the production of Black wealth or even Black survival.

The YOUR project does not directly address the issue of systemic racism in the urban environment. But there is a way to read its appearance in the city as gesturally linked to the ongoing structural racism plaguing Black and brown communities everywhere. This reading starts with the medium of its intervention: that is, a body scaling the walls of a piece of private property to illegally alter its surfaces. An act of property crime, graffiti writing in this context cannot be considered without a larger understanding of racialized police violence in the service of private property.

In Miami Beach, one widely publicized case helps to illustrate this point. In August of 2013, Israel Hernandez, a young artist who worked under the moniker "Reefa," was murdered by police when he was caught tagging a vacant building. Eighteen-year-old Hernandez died of a heart attack after the police shot him with a taser gun. His death made national news in a still-ongoing period in which police violence against people of color frequently makes headlines. His death sparked protests and debates across Miami and the country. Hernandez became Miami's face of the Black Lives Matter movement. His name, listed alongside those of Eric Garner, Mike Brown, and more recent additions of Breonna Taylor and George Floyd, along with many others, positioned the often politically disengaged landscape of the city within the context of national struggles for racial justice. While we have no way of knowing the race, gender, or class position of the YOUR's author, we do know that participating in illegal graffiti writing entangles artists in the racialized politics of policing in the city. The risk the author took in scaling the hospital's walls and painting her message would be greater or lesser depending on the color of her skin. For Black or brown artists, the consequence of such an action may be fatal.[44]

In addition to highlighting the horror of systemic racisim, Hernandez's murder also indicated the extent to which notions of value in the urban landscape are filtered through the global regime of capitalism. Hernandez was killed for a minor property crime. That police, tasked with protecting the rights of owners rather than those of the users of the city, exercised violence in their pursuit only indicates the terrible costs of the current system. Whether the properties thus protected will soon be underwater is of little relevance to Hernandez or the people who cared for him. For them and so, so many other Black and brown people who have suffered police violence, the catastrophe is not coming. It is already here. It has been here for a very long time. Cases like Hernandez' make clear how crucial Harvey's "right to the city" is. Hernandez' death is not simply coincidental with "accumulation by dispossession." Given that Miami Beach's development has effectively driven impoverished and middle class workers entirely off the island, those million dollar homes have become structures to be protected, apparently at any cost, from those to whom the city no longer seems to belong. Indeed, their survival is sometimes categorically denied if it threatens even so much as an abandoned building: even the chance to demonstrate resilience, it seems, is not equitably distributed. Miami police and its property owners claim the right to the city under the rubric of resilience even as their very public policies, governance, and policing mechanism deny the right to the city to some among the community. Developers and property owners demand resilience from city-dwellers even as they prevent some from surviving at all. In this way the right to the city is effectively wrested from its proper place in the commons–which includes Hernandez and the anonymous graffiti writer of YOUR, among others.

The intervention of YOUR into the specific landscape of Miami Beach highlights the precarity of the graffiti writer: a precarity that is physical, economic, and social. The writer is imperiled by the athletic feat of the work itself: scaling the top floors of a building without supporting infrastructure and in the dark is certainly a treacherous endeavor. The writer is potentially in an endangered position within the complex of racialized development and policing processes. And she very likely cannot afford to live anywhere near the site of her work. Million dollar homes may be in a precarious position vis a vis the rising seas, but there are much more valuable bodies, human bodies, at risk.

Fantasy land, plastic city

Precarious bodies and communities in Miami are not symmetrically distributed across the landscape, nor are the consequences of catastrophes wrought by climate change or urban development. It is here that the speculative fantasy of YOUR is most visible.

A recent Harvard study, focused specifically on the case of Miami-Dade County, describes climate gentrification as an increasing risk to urban life. The authors of the study write that

> climate change impacts arguably make some property more or less valuable by virtue of its capacity to accommodate a certain density of human settlement and its associated infrastructure. The implication is that the price volatility associated with rent seeking, speculative investment, or superior purchasing

> power is either a primary or a partial driver of the patterns of urban development that lead to displacement (and sometimes entrenchment) of existing populations consistent with conventional framings of gentrification.[45]

In layman's terms: climate change, or the perception of climate change, will likely contribute to the continued displacement of lower-income communities across Miami's urban landscape. In a thorough examination of the increase in property values across the county from 1971 to 2017, the study's authors found that elevation already had a distinct impact. The higher the land, the more likely an increase in the property value. The same study also found decreases in property values likely related to "nuisance flooding" in the typically wealthy coastal areas of Miami Beach and Biscayne Bay.

Urban studies scholar Richard Florida, in his review of the study, writes,

> In Miami, the wealthy have long preferred the coasts. But as the risk of climate change grows, this will likely change, with the wealthy colonizing the higher, less flood-prone ground inland and especially in and around downtown. Indeed, as the study shows, it is the higher places—traditionally home to the less advantaged and the poor—that have seen the largest jumps in price appreciation.[46]

In this case, the vertical movement is literally a movement to displace and dispossess. Both upward moving wealth and the expectation of endlessly rising profits work together to shape the city in ways that benefit some and disadvantage others.

Given these conclusions, YOUR's prediction requires an addendum. The million dollar homes on Miami Beach and elsewhere in the high-end coastal communities of South Florida will soon be underwater. But by then their owners will most likely have already pushed out low- and middle-income families on higher ground elsewhere in the city. Million dollar condos will soon be going up in communities like the mostly-Black and low-income neighborhoods of Liberty City and Overtown, both of which are further inland and on considerably higher ground.

The fantasy projected by YOUR, that climate change might level the economic playing field, or at least deliver to the rich the same kinds of damage it delivers to the poor, reveals not only the kinds of tensions that climate change will continue to both highlight and exacerbate within the urban terrain, but also, and more importantly, that fantasies about what the city can be *shape* the city as it exists now, sometimes quite literally. Fantasies themselves have power to both see the city as it is and project speculatively forward in the hopes of altering the city's course in the future.

YOUR responds, in part, to the epistemological problem, described by Horn and Nixon, of the catastrophic impacts of climate change. Though its consequences are already being felt across South Florida, they are only occasionally experienced as singular events, as "weather." This leads to gaps in representation and a dearth of robust views of the broad swath of urban life impacted by climate change; it leads to difficulties exposing affective and social worlds shaped by a warming world. Computational models that indicate a coming catastrophe cannot work to invite in these other registers. But YOUR's dark prophecy positions the city in relation to exactly such worlds, especially if the full context of its intervention is addressed: its performance, its fantasy, its material and historical context within the region.

Horn writes that "global warming may be the most threatening symptom of the Anthropocene, yet it is also one particularly complicated epistemically, given the highly abstract concept of climate operative today. How can we conceive of a threat that is so ungraspable and so vast that it eludes the scales of human experience?"[47] One answer, proposed by YOUR and the other projects addressed in this book, is to produce works of speculative play that intervene in the material landscape of the city to help expose its forms (physical, social, and economic) in order to reshape them. In the work of YOUR, there is more than a fantasy about a more equitable, post-catastrophic future. By the very nature of its medium, YOUR offers a concrete understanding of the city as *alterable*. It declares that the city as it is does not have to be, and indeed will not be, the city of the future. Both its message and its mechanism of delivery make the point that the city can and does change. And in this sense, YOUR understands the city not as resilient and flexible, but as plastic.

My claim that YOUR understands the city as plastic goes beyond the basic understanding of a city as always in flux. Of course, the appeal of cities is the lively way they transform, from moment to moment as one wanders around, from year to year as the built environment ages, transforms, shifts. Calling the city plastic, however, suggests something more. Plasticity reimagines much more radical movement; it points out that within the urban system there remain many possible exploits, and many possible futures, catastrophic, utopic, and otherwise.

Catherine Malabou has written extensively about the concept of plasticity from Hegel on. In her book *What Should We Do with Our Brain?* she considers brain plasticity as a radical opening to change the way we can live. I use her discussion of it here to underline the clear ways in which, in her philosophy and my own reading of Miami's urban landscape, plasticity indicates a material capacity to both receive and give form (i.e., brain plasticity) and to refuse form, to deflagrate (i.e., plastic explosives). Malabou writes:

> Meanwhile, plasticity directly contradicts rigidity. It is its exact antonym. In ordinary speech it designates suppleness, a faculty for adaptation, the ability to evolve. According to its etymology–from the Greek *plassein,* to mold–the word *plasticity* has two basic senses: it means at once the capacity to *receive form* (clay is called "plastic" for example) and the capacity to *give form* (as in the plastic arts or plastic surgery). [...] But it must be remarked that plasticity is also the capacity to annihilate the very form it is able to receive or create. [...] We thus note that plasticity is situated between two extremes: on the one side the sensible image of taking form (sculpture or plastic objects) and on the other side the annihilation of all form (explosion).[48]

The brain is plastic in that it is and will be altered always in relation to the world. Malabou traces recent developments in neuroscience to make the case that we have moved from an understanding of the brain that indicates top-down governance, the brain as responder to the world to, instead, a world-making, processual brain. More: plasticity allows the brain to change and reconfigure *itself*. The brain is plastic in that it can alter and will be altered, materially, and that such alterations take place always in relation to each other and the world. The city, too, is plastic in exactly this way. I want to be careful to avoid over-metaphorization here: it is not that the city is like a brain. It is that the characteristics of plasticity, as Malabou traces them, are also those of the city. Plasticity may be, in fact, the defining characteristic of the city, and one that opens up for us the best possible hope for a different city.

Plasticity, as a descriptor of the urban world and of acts that appear and disappear within it (as in YOUR), is also exceptionally functional in its opposition to flexibility. Flexibility is the ideal, subjected material body of city or worker under the current biopolitical regime. Where urban developers and investors imagine the city as flexible, YOUR and projects like it understand it as plastic. Where resilience discourse relies on flexibility (flexible business models and zoning ordinances, flexible populations who must always be ready to adapt), it omits and elides the material fact of urban plasticity.

This distinction is an essential one. Malabou describes flexibility as "the ideological avatar of plasticity – at once its mask, its diversion, and its confiscation."[49] The work of graffiti writing is interventional, it materially refuses flexibility in a number of ways, not least of which, at least in the case of YOUR, by directly attacking the kind of development practices and property rights that so often depend upon neoliberal economic models, in which flexibility (touted *as* resilience in some cases) is not only an economic, but a moral good. Resilience discourse in the city heavily relies on notions of flexible populations and urban plans: but it often works *against* fundamental, radical change. In the face of rising ocean waters, the flexible city bends to climate gentrification. YOUR disturbs the ideological claims to flexibility by employing a sense, quite closely akin to that described by Malabou, of the city as plastic. It indicates in its performance and even its fantasy, that the city could be and is the site of explosive transformative possibilities. YOUR is, after all, both a destructive act and a call for a certain explosive reshaping of urban life in the face of catastrophe.

YOUR has a particular temporal framing of catastrophe as well: it is teleological. The ecological disaster (already very present along Florida's coastline) is a definitive end, necessarily coming, that it fantasizes will level the economic playing field by destroying a city this piece of illicit public art sees as having already mired itself in catastrophic (in both the ecological and economic senses) development practices.

YOUR understands the urban ecology as fragile. It also understands that this fragility is in part produced by and through economic traffic. This fragility of the urban ecology that is so central to this particular piece of public art is put in conversation with another, more immediate and material fragility: that of the body/ies of the work's producer(s). The artist's body is here quietly, invisibly put at great risk, its own precarity mobilized to mark the precarity of urban life in the Miami of the present. Because this work understands the city as plastic, however, and because it intervenes in that plastic city with a speculative declaration, YOUR undercuts and destabilizes its own teleological line. In the act of claiming a coming catastrophe, it indicates that the city is radically alterable and thus that such catastrophic futures need not inevitably unfold.

Ultimately, the only way to change the city, to build urban space that is not so totally governed and policed by the laws of capital accumulation, is to insist on plasticity as the key fact of the urban landscape. It is only within this context that the epistemic challenge of climate change, or capital itself (unrepresentable as it is, according to Jameson) can be harnessed for just causes. Transformative resistance can only occur in a plastic city.

I would like to return, here, to Harvey's "right to the city." His vision of what that right is, and what it should be, is also essentially plastic:

> The right to the city is far more than the individual liberty to access urban resources: it is a right to change ourselves by changing the city. It is, moreover, a common rather than an individual right since this transformation inevitably depends upon the exercise of a collective power to reshape the processes of urbanization. The freedom to make and remake our cities and ourselves is, I want to argue, one of the most precious yet most neglected of our human rights.[50]

That freedom, to make and, (essentially) remake the city, is a freedom dependent upon plasticity in that it can only be claimed when we, collectively, acknowledge that the city is *not* categorically and universally resilient, even if it is adaptable. We must go further and insist that flexibility under the rubric of resilience fails to address the actual alterability of our urban worlds. When Malabou uses the phrase "our brain" she is writing of a common coming into being. Our city, like our brain, is plastic: it can be reformed, take on new forms, explode forms. A resilient city is the plastic city's dark, neoliberal mirror and its veil. When the hospital building was demolished to make room for a new luxury condo complex, the developers' success was partly due to their reliance on resilience and on flexibility to hide the common nature of the right to the city and to maintain the hierarchies that wrest it from the common.

Sooner or later we will have to decide, in the face of both ongoing and coming catastrophes, what city it is we want to build, together, with our plastic brain. Harvey insists that the right to the city is necessarily a *common* rather than individual right. If we can come to know the plastic city, we will have the opportunity to recognize our right to the city, we will have a chance to demand it and, in so doing, remake our world.

That does not mean the flattening of all differences, particularly in a city envisioned as plastic. Malabou suggests that antagonisms, the dividing between certain "us's" and certain "thems" may be an essential step toward refusal, to changing ourselves and our cities with our plastic brains. Antagonism is, after all, oppositional to docility. YOUR's declaration in the city is antagonistic in the way it divides a speaking self and another delineated by its (potentially precarious) class position. That the statement is in English also delineates a certain audience. Miami's native language might be said to be Spanish, and Miami Beach, with its constant influx of tourists and immigrant workers, is wildly linguistically diverse. In this light, it is possible to read the intervention as working directly against the common nature of the right to the city, even of Malabou's "our brain." But I want to read against that grain here. The common right to the city does not mean a refusal of divisions, even antagonisms: confrontation and conflict are not mutually exclusive of the exercise of the right to the city. From Lefebvre on, the right to the city has been seen as commonly shared, but not as requiring total coherence. It is held by a mass of urban folk who can come together despite their differences against the very small segment of the population, the private and corporate interests that currently monopolize the city and its development policies. If YOUR delineates a "you" against which it positions itself, it is these interests, the small segment of the population who owns the city. If YOUR is itself a way of calling into being a subject, it is a political subject whose right to the city is currently denied. She is a class-conscious subject who indeed positions herself against those who would deny that the city belongs to those who use it. Harvey has noted that "increasingly, we see the right to the city falling into the hands of private or quasi-private interests."[51] YOUR is situated directly against those interests: it is an illegal act targeted directly at private property in the process of capitalist redevelopment. It also calls out the flatness and smoothness of resilience discourse by operating so distinctly within the context of Miami Beach, whose developers and municipal governance constantly touts resilient and flexible economic models, while simultaneously refusing to so much as look at the actual needs of city users, thus building the very barriers to resilience it is supposed to be addressing.

Malabou writes:

> To cancel the fluxes, to lower the self-controlling guard, to accept exploding from time to time: this is what we should do with our brain. It is time to remember that some explosions are not in fact terrorist — explosions of rage, for example. Perhaps we ought to relearn how to enrage ourselves, to explode against a certain culture of docility, of amenity, of the effacement of all conflict even as we live in a state of permanent war. It is not because the struggle has changed form, it is not because it is no longer possible to fight a boss, owner, or father that there is no struggle to wage against exploitation. To ask "what should we do with our brain?" is above all to visualize the possibility of saying no to an afflicting economic, political, and mediatic culture that celebrates only the triumph of flexibility, blessing obedient individuals who have no greater merit than that of knowing how to bow their heads with a smile.[52]

The culture of docility of which she speaks can easily be seen in resilience discourse's perpetual appeal to flexibility and adaptability and its refusal of structural critique. YOUR does not accept the neoliberal fiction of a universal community where we are all "flexible" enough to get along. It is an explosion of rage at an economic system that places some in more vulnerable positions than others while simultaneously refusing our common vulnerability. YOUR's antagonism is essential to its radicality: it picks sides. Ultimately, plasticity is a *capacity*, not a quality of being. In this, it works directly against resilience, which is consistently framed as an essential quality to be adopted and displayed by cities, communities, and individuals. Plasticity engenders agents rather than audiences. It opens up ways to act.

You can now, of course, buy T-shirts, mugs, even duvet covers with the "slogan" first written across the abandoned hospital. And here emerges, as with each of the central projects discussed in this book, the question of the site of urban resistance and its potency. In a world in which what might appear as radical interventions can be used for the continued production of consumer goods, and in which urbanization has been (as Harvey, among others, has effectively argued) so tightly tied to capital accumulation, we must wonder when and whether urban work like that discussed here can change the city, can stave off the coming catastrophe, can disrupt the catastrophe now ongoing. I believe it can and does, but we must acknowledge that the insights of these interventions do not pull them safely outside the realm of contemporary capitalism. For this reason they must be, like YOUR, intentionally plastic. Acknowledging, underlining, and intervening in the plastic city will always be a process. It will require patterned repetitions of refusal. For each novelty T-shirt that emerges from the site of an intervention, there must be still more resistant, explosive, playful, and creative engagements moving forward.

Proximity and vulnerability: Little Haiti and disastrous discourse

I cannot conclude this case study, based as it is in Miami, without acknowledging the less visible catastrophe that links the city, and the other cities that are the sites of case studies in this book, to a certain kind of racialized discourse that disavows these cities' strong links to communities elsewhere. In Miami's case, that elsewhere is very often Haiti.

Despite proudly claiming itself as a diasporic and immigrant city, Miami and, as Kaiama L. Glover has argued, the bulk of the developed world have clung to "a narrative patchwork" that helps to form "a comfortable quilt – a security blanket shielding the First World from its proximate others."[53] She writes that

> to think about Haiti is to think about the whole of the Atlantic world. Haiti has long been the beating heart of an intricate corpus of narratives, ideological setups, forms of visual culture, and affective practices that are deeply and enduringly racialized. Confirmed site of disastrous not-quite-humanity, Haiti is ur-example of the Afroabject.[54]

Positioning Haiti as that ur-example allows for perpetually repeating the fantastic claims, sometimes explicit, that Black subjects are predisposed to catastrophe and that the long-running campaign of intervention by Euro-U.S. powers is blameless in Haiti's ongoing suffering.

In Miami's case, such claims are rendered even more radically absurd in that the city's development and its current landscape are in fact tightly tied to at least some of these interventions. Glover cites a long list of policies employed by the United

States and Europe that have perpetuated both political and economic disaster in Haiti, including "the Central Intelligence Agency's support of the Duvalier regimes of the 1970s and 1980s, during which time Haiti served as a military bulwark against Fidel Castro's communism while its leaders' brutal and corrupt governance was underwritten by the United States."[55] These strategies brought a huge number of Haitian refugees to Miami and embroiled members of the Cuban diaspora in the city in an antagonistic relationship with its Haitian neighbors. The majority of Haitians who arrived in the United States did so before the 2010 earthquake, fleeing the violence of their home country. Today, Miami's Haitian community is the nation's largest.

The city's relationship to Little Haiti is a microcosm, in many ways, of the white, European and North American worlds' relationship to Haiti. In boosterist materials, the city happily claims the neighborhood as a beacon of Miami's diversity, even as it consistently refuses to offer legitimate protections or services to its inhabitants. One recent and typical example is the municipal approval, despite complaints from longtime neighborhood residents, of the "Magic City Innovation District." As the *Miami New Times* put it:

> The proposed Magic City Innovation District would instantly transform Little Haiti from an area of Caribbean immigrants and locally owned shops into a glitzy, landlocked version of Miami Beach. Though Little Haiti is already gentrifying at a rapid clip, locals worry the development would guarantee that the last stretch of affordable housing in the area would vanish.[56]

The article goes on to point to several studies that have placed Little Haiti among the fastest gentrifying areas in the country, a status no doubt related to the fact that the community sits on some of the city's highest ground. Unsurprisingly, the plan adapts the language of efficiency, flexibility, and resilience that is now all but mandatory in development projects. It claims to offer "forward thinking urban planning and consciousness design [...] in order to minimize environmental impact, ensure resiliency, maximize efficiencies to reduce long term operating costs and to promote a healthy environment for people to live, work, learn and play."[57]

The proposal is scant on on concrete details of how local residents will actually reap the benefits of this development. Indeed, as of this writing, one Little Haiti resident, prevented from formally objecting to the development plan, is suing the city. The proposal focuses on economic growth, rather than social welfare. It describes the neighborhood as one of "incredible growth potential" that "is attracting Miami's creative class and urban enthusiasts."[58] While an attached foundation promises aid to the local community in the form of job training and cultural preservation, with no clear numbers available it is difficult to imagine that the planned district and its luxury condo development won't drive up prices and displace local residents. In fact, it already has: the Magic City Trailer Park, an 86-year-old mobile home community housing 40 working-class families, was forced out to make way for the development project.[59]

While there is little doubt that municipal governments and real estate developers in the city would be happy to see gentrification and condo construction city-wide, they have been considerably less apt to rush so quickly through similar development proposals in Little Havana, for example, or in majority white Coral Gables. Haitians immigrants to the United States and elsewhere have suffered decades of ingrained racism and Miami is no exception, despite being a city literally founded on the labor of Afro-Caribbean bodies.[60]

One can look at the national stage as well to see the perpetuation of a racialized discourse vis a vis Haiti. In an infamous recent example, President Donald Trump referred to Haiti as a "shithole," and his administration is fighting to deny the continuation of temporary protected status granted to Haitian immigrants after the massive earthquake, despite very clear evidence that their livelihoods and safety remain imperiled in their home country. All of this plays out in acute ways in Miami's Little Haiti, and in the ongoing tensions between the Hatian community and both the Latino and Anglo forces who generally control local policy.

But more than this, of course, is the much larger scale, historically rooted problem to which Glover points: one in which many projects seeking to aid Haitian communities are actually participating in the production of a racialized disaster porn meant to ease the consciences of white Westerners: "The drive to consume disaster and the implication that disaster is an ontological reality for the other is [...] a comfort."[61]Such comfort will, sooner or later, be a cold one. If catastrophe is imagined as always happening to other, and further, is seen as her ontological reality, then in addition to perpetuating widespread inequality, this racialized discourse also perpetuates both a denial of our shared vulnerability, and an ongoing deferral and disavowal of catastrophe.

Glover argues that, beyond the mapping of difference that one might see in a military map, locating borders around states and oceans between islands, the white Western public's obsession with Haiti as the site of brown and Black catastrophes is its own kind of border production, "meant to allay collective 'One Percent' anxieties regarding an imminent 'becoming abject.'"[62] There is a kind of disaster relief that positions Haiti, and other sites of the Afro-abject, as a kind of pornography that white audiences consume to allay their fears and obscure the fact that they share positions of precarity and mortality with Black others. Whatever risks Haiti faces as the impacts of anthropogenic climate change and global capitalism make themselves every year more visible will not be isolated to Black and brown bodies on islands near or far. Though those bodies will almost certainly suffer disproportionately, the here and now of Miami Beach is not immune. In this sense, the graffiti writer is quite right to claim at least a minimal sort of leveling the field. YOUR also, if minorly, inverts the othering logic of racialized discourse regarding the Afro-abject (always framed as poor). Instead it is the white millionaire who will be the subject of the coming catastrophe. The "your" to which the sign points is, in its fantasy of a future Miami, all but guaranteed to be white. As one study of Miami conducted by the Corporation for Enterprise Development found:

> Haitians and African Americans are more likely than any other major ethnic groups to live in poverty, with poverty rates of 45% and 44%, respectively.

> While median household income among Miami's Latino residents varies from $46,000 among Mexicans to just over $25,000 among Cubans, African American and Haitian household incomes hovers just above $21,000. (In contrast, the median white household income is more than three times greater, at over $70,000.)[63]

But, as I have argued, YOUR elides through fantasy the disproportional nature of the suffering perpetuated by global capitalism and, in Miami's specific urban landscape, the attendant climate gentrification. That Little Haiti is consistently pinpointed as among the most rapidly gentrifying neighborhoods in the United States is, yes, partly because it stands on higher ground. But, as the Magic City Innovation development's progress and a long history of racism in the city suggests, it is also because Little Haiti has been historically occupied by Black Haitian immigrants whom the city and the wealthy property-developers it supports and protects do not see as valuable. Happy to consume Haiti as the site of disaster porn, white developers and investors refuse to acknowledge their complicity in producing disasters for local Haitian communities. The Magic City Innovation District all but ensures that many Haitians will be pushed out to be forgotten in other, more distant suburbs of Miami where their culture and community may fragment.

Thinking about the relationship Miami shares with its own Haitian community, and the ideological position of Haiti in Euro-U.S. discourse in a book about how catastrophe is imagined and played out in the urban landscape has the critical capacity to do two things. First, it works to destabilize the construction of Haiti as "ur-other," a construction that should not be possible be in a city so physically close to Haiti and so full of that island's vibrant cultural gifts. Second, it underlines that in contending with catastrophe, or, perhaps especially, imagining it, one must take extreme care not to engage in what Glover calls the "drive to consume disaster." If YOUR is a fantastic imagination of the comeuppance on its way for Miami's wealthiest in the face of a climate catastrophe, it is, too, a gestural reminder of the ways that borders of all sorts, ideological, material, military, fail. As Glover writes, "the "problem" is that borders insufficiently contain abjection. They bring that which they divide into the very closest proximity."[64]

The concept of resilience is deeply rooted in racist and xenophobic narratives about national security, as Thomas argues in her book. This of course means that it is closely tied to the production of borders. Its very use as an appeal to flexibility, as a mechanism meant to inculcate and shape Black bodies into good neoliberal subjects who pop back into shape after suffering some kind of catastrophic blow, effectively evacuates the possibilities of resistance. In the case of development projects like the proposed Magic City Innovation District, the steamrolling of Hatian-American voices and refusal to address their concerns is one more instance in which Black, brown, and immigrant subjects are not seen as responding "appropriately" within the urban context. The projects that parrot and advocate resilience leverage it as a weapon against these subjects' rights to the city. Plasticity, as a formal refusal of resilience discourses' insistence on flexiblity, must be underlined as the essential nature of the city because to position the city as plastic is to allow for change in ways other formulations of the urban world do not. The uprisings which grew in response to the police murder of George Floyd that have unfolded across the United States (including Miami) and the world are making abundantly clear that plasticity, which can embody both explosion and transformation, was *already* the real form of the city. It was urban plasticity that engendered the very conditions for protest. Plasticity has always, in fact, been the underlining form of urban life that makes it a site for productive revolt. The vertical formation against which the uprising stands, white supremacy (so closely tied with the history of urban development in the United States) and its valuation of white bodies and white lives both above and at the expense of Black and brown lives, can only be toppled in a plastic world.

The catastrophe that is racialized discourse has dire consequences in Miami. The Haitian, Afro-Caribbean, and Black populations who helped to build the city and continue to contribute to its vibrancy are largely ignored under the city and the country's current political regimes until being offered up as a spectacle of catastrophe. They are excluded even by some of the region's so-called resilience strategies. Such discourse, embodying as it does the deferral of catastrophe and disavowal of responsibility, denies the real links between communities: between Haiti and the United States, between local members of the Haitian diaspora and other communities in Miami, between the rich home-owners of Miami Beach and the poor residents of Overtown, or Liberty City. The reality is that all these groups have something in common: the waters are rising and it is our homes, our communities, our public spaces at risk. Perhaps, in place of a fantasy that climate change will decimate the homes of the capitalist elite, we might instead begin to imagine ways in which our particular affinities could bind us against the powers that oppress us all.

Notes

1 Photo courtesy of Stephen Cohen.
2 Tristram Korten, "In Florida, Officials Ban Term 'Climate Change,'" *The Miami Herald, (FL)*, March 8, 2015, https://www.miamiherald.com/news/state/florida/article12983720.html.
3 T.D. Allman, *Finding Florida: The True Story of the Sunshine State* (New York: Atlantic Monthly Press, 2013).
4 Ashley Dawson, *Extreme Cities: The Peril and Promise of Urban Life in the Age of Climate Change* (London: Verso Books, 2017), 18.
5 Dawson, *Extreme Cities*, 18.
6 USGCRP, "Fourth National Climate Assessment," 2018, https://nca2018.globalchange.govhttps://nca2018.globalchange.gov/chapter/19.
7 Jan Nijman, *Miami: Mistress of the Americas* (Philadelphia: University of Pennsylvania Press, 2011).
8 Nijman, *Miami*, 201.
9 Nijman, *Miami*, 109.
10 Parks, *Rethinking Media Coverage*, 9.

11 Brickell, a neighborhood in downtown Miami and the home to many of the city's newest luxury condo units saw the waters rise to three feet in the storm-surge caused by Hurricane Irma, turning its main drag into a rushing river under gale winds. "Irma Hit Downtown Miami — and Turned Its Biggest Streets into Rivers," miamiherald, accessed December 17, 2019, https://www.miamiherald.com/news/weather/hurricane/article172495761.html.
12 Eva Horn, *The Future as Catastrophe: Imagining Disaster in the Modern Age*, trans. Valentine Pakis (New York: Columbia University Press, 2018), 40.
13 Horn, *The Future as Catastrophe*, 56.
14 Horn, *The Future as Catastrophe*, 45.
15 Horn, *The Future as Catastrophe*, 43. Emphasis original.
16 Horn, *The Future as Catastrophe*, 43.
17 See my discussion of these two forms of speculation (via the work of an uncertain commons) in Chapter 1.
18 Lindsay Thomas, "Forms of Duration: Preparedness, The Mars Trilogy, and the Management of Climate Change," *American Literature* 88, no. 1 (2016): 159–84.
19 an uncertain commons, *Speculate This!*, n.p.
20 Thomas Sparrow, "Latin America's Internet 'Brain,'" *BBC News*, January 31, 2013, sec. Technology, https://www.bbc.com/news/technology-21178983.
21 See Rene Rodriguez, "Miami's Housing Affordability Crisis Poses a Dire Threat to the City's Future, Experts Say," *The Miami Herald*, October 4, 2019, https://www.miamiherald.com/news/business/real-estate-news/article235758107.html.
22 Dawson, *Extreme Cities*, 13.
23 Dan Weissmann, "Miami Condo-Buyers Aren't Homeowners. They're Traders.," *Marketplace*, February 10, 2015, https://www.marketplace.org/2015/02/10/sustainability/water-high-price-cheap/miami-condo-buyers-arent-homeowners-theyre-traders.
24 David Harvey's exhaustive look at the cyclical nature of catastrophe in 20th and 21st century capitalism makes it more than clear that, without substantive changes in the way capitalism operates, another economic crisis, even catastrophe is all but inevitable. See David Harvey, *The Enigma of Capital and the Crises of Capitalism* (New York: Oxford University Press, 2010).
25 Gary Rivlin, "Why New Orleans's Black Residents Are Still Underwater After Katrina," *The New York Times*, August 18, 2015, sec. Magazine, https://www.nytimes.com/2015/08/23/magazine/why-new-orleans-black-residents-are-still-under-water-after-katrina.html.
26 Jeremy Walker and Melinda Cooper, "Genealogies of Resilience: From Systems Ecology to the Political Economy of Crisis Adaptation," *Security Dialogue* 42, no. 2 (2011): 143–60.
27 Walker and Cooper, "Geneologies of Resilience," 54.
28 "Urban Resilience," 100 Resilient Cities, accessed December 17, 2019, https://www.100resilientcities.org/our-impact/
29 "Urban Resilience," http://www.100resilientcities.org/resources/
30 "resilience, n.". OED Online. September 2020. Oxford University Press. https://www.oed.com/view/Entry/163619?redirectedFrom=resilience (accessed September 17, 2020).
31 Lindsay Thomas, *Training for Catastrophe: Fictions of National Security After 9/11* (Minneapolis: University of Minnesota Press, forthcoming).
32 "Resilient305" (Greater Miami and the Beaches, 2019), http://www.mbrisingabove.com/wp-content/uploads/Resilient305_final.pdf.
33 "resilience, n.". OED Online. June 2020. Oxford University Press. https://oed.com/view/Entry/163619?redirectedFrom=resilience (accessed June 10, 2020).
34 Kyra Gurney, "Miami Beach Plans for a Park, High-Rise at South Beach Gateway Move Forward," *Miami Herald, The (FL)*, August 5, 2018, Access World News – Historical and Current. https://www.miamiherald.com/news/local/community/miami-dade/miami-beach/article215776370.html

35 Gurney, "Miami Beach Plans," https://www.miamiherald.com/news/local/community/miami-dade/miami-beach/article215776370.html.

36 Video still, courtesy *The Miami Herald*. https://www.miamiherald.com/news/local/community/miami-dade/miami-beach/article229323424.html

37 David Harvey, "The Right to the City," *New Left Review*, no. 53 (October 2008): 23–40.

38 Harvey, "The Right to the City," 34.

39 Josh Baumgard, "19 Miami Metro Cities Ranked by Median Rent Prices," *Curbed Miami*, December 20, 2017, https://miami.curbed.com/2017/12/20/16801052/miami-rent-report-boca-beach.

40 Kevin T. Greiner, "The Dynamics of Housing Affordability in Miami-Dade County" (Miami: South Florida Housing Studies Consortium, April 2017), https://civic.miami.edu/_assets/pdf/housing-initiatives/housing-reports/Dynamics-of-Housing-Affordability-Inclusionary-Zoning-2017-4-19-Final.pdf.

41 Alejandro Portes and Ariel C. Armony, *The Global Edge: Miami in the Twenty-First Century* (Berkeley: University of California Press, 2018), 33-34.

42 "Resilient 305" 2019.

43 See the University of Miami's Housing Solution Lab's Miami Housing Policy Timeline, "Changing Neighborhoods," accessed June 8, 2020, http://www.cdn.miami.edu/wda/cce/Documents/Miami-Housing-Solutions-Lab/changingNeighborhoods.html.

44 The American Civil Liberties Union conducted a study of data on all adult criminal defendants in Miami-Dade County from 2010 to 2015. They found "disparities at every decision point that, regardless of ethnicity, result in disadvantages for Black defendants and neighborhoods while resulting in advantages for White defendants and neighborhoods." While the report did not address police violence of the sort that killed Hernandez, it is more than clear that Miami-Dade's criminal justice system, from policing through trial and sentencing, strongly disadvantages Black Miamians. See: ACLU of Florida and Greater Miami, "Unequal Treatment: Racial and Ethnic Disparities in Miami-Dade Criminal Justice", July 20, 2018. https://www.aclufl.org/sites/default/files/aclufl_unequaltreatmentreport2018.pdf

45 Jesse M. Keenan, Thomas Hill, and Anurag Gumber, "Climate Gentrification: From Theory to Empiricism in Miami-Dade County, Florida," *Environmental Research Letters* 13, no. 5 (April 23, 2018), https://iopscience.iop.org/article/10.1088/1748-9326/aabb32.

46 Richard Florida, "'Climate Gentrification' Will Deepen Urban Inequality," *Citylab* (blog), July 5, 2018, https://www.citylab.com/equity/2018/07/the-reality-of-climate-gentrification/564152/.

47 Horn, *The Future as Catastrophe*, 63.

48 Catherine Malabou, *What Should We Do With Our Brain?*, trans. Sebastian Rand (New York: Fordham University Press, 2008), 5.

49 Malabou, *What Should We Do*, 12.

50 Harvey, "The Right to the City," 23.

51 Harvey, "The Right to the City," 38.

52 Malabou, *What Should We Do*, 79.

53 Kaiama L. Glover, "'Flesh Like One's Own': Benign Denials of Legitimate Complaint," *Public Culture* 29, no. 2 (82) (May 1, 2017): 235–60, 237.

54 Glover, "'Flesh Like One's Own'," 236.

55 Glover, "'Flesh Like One's Own'," 240.

56 Jerry Iannelli, "Miami Little Haiti 'Magic City Innovation District' Slammed by Activists" *Miami New Times*, September 26, 2018, https://www.miaminewtimes.com/news/miami-little-haiti-magic-city-innovation-district-slammed-by-activists-10767264.

57 "Inside Magic City," *Magic City Innovation District® - Miami's Innovation and Entertainment District* (blog), accessed June 9, 2020, https://magiccitydistrict.com/inside-magic-city/.

58 "Inside Magic City," https://magiccitydistrict.com/neighborhood/#history
59 Erik Bojnasky, "The Passing of a Neighborhood," *Biscayne Times*, April 2015, http://biscaynetimes.com/index.php?option=com_content&view=article&id=2051:the-passing-of-a-neighborhood&catid=50:community-news&Itemid=258.
60 For an exhaustive study and historical overview of the case of Miami in particular, see "The Color of Wealth in Miami Metro," The Kirwan Institute for the Study of Race and Ethnicity at The Ohio State University, February 2019, http://kirwaninstitute.osu.edu/wp-content/uploads/2019/02/The-Color-of-Wealth-in-Miami-Metro.pdf.
61 Glover, "'Flesh Like One's Own'," 248.
62 Glover, "'Flesh Like One's Own'," 249.
63 "The Racial Wealth Divide in Miami" (Prosperity Now, October 2016), https://prosperitynow.org/sites/default/files/resources/Racial_Wealth_Divide_in_Miami_RWDI.pdf.
64 Glover, "'Flesh Like One's Own'," 249.

Interlude 2

Plastic/city

Roland Barthes, ruminating on plastic in the 1950s, saw the substance as a marker of a new relationship between people and nature. He noted that plastic's qualities seemed less connected to objects and more to movement. As an indicator of capitalism's transformation of life, plastic, like the commodity in Marx, presented itself as magical. Its invention and proliferation in the 20th century indicated, for Barthes, a movement away from the impulse to imitate that which is rare in nature: plastic reorders the hierarchy of forms, and the everyday and the functional begin to dominate aesthetics:

> Plastic is wholly swallowed up in the fact of being used: ultimately, objects will be invented for the sole pleasure of using them. The hierarchy of substances is abolished: a single one replaces them all: the whole world can be plasticized, and even life itself since, we are told, they are beginning to make plastic aortas.[1]

Barthes might have been prescient in his reading of plastic, though the plasticization of the world, of life itself, went in directions he could not yet imagine in its shaping of and imbrication with the natural. For one, the ubiquity of disposable plastics has had dire ecological consequences, perhaps the most drastic example of which is the Great Pacific Garbage Patch. (One model, published in *Nature* estimated at least 79,000 tons of plastic floating atop the ocean in an area of 1.6 million square kilometers[2]). Plastic as a material could thus be characterized not just as an upheaval of a once sacred hierarchy of forms but as a kind of catastrophe.

Plastic is also exemplary of the strong and material entangling of nature and culture already indicated in Barthes. As Dana Philips succinctly put it: "Nature is thoroughly implicated in culture and culture is thoroughly implicated in nature."[3] But the quality Barthes found so compelling about plastic, that it is "the very idea of infinite transformation," is one it shares with the city.[4] And, like plastic itself, the plastic city has incredible potential for transformation and carries with it, always, the threat of catastrophe: ecological, social, and economic.

You will recall from Chapter 3 that I use "plasticity" as it has been elaborated by Catherine Malabou, focusing on her designation of it as "a faculty for adaptation, the ability to evolve"; the same quality outlined by Barthes. Plasticity is both a capacity for giving and receiving form and, essentially, "the capacity to annihilate the very form it is able to receive or create."[5] I argue that plasticity has a particularly potent force in the urban milieu and offers a more accurate account of the material and social realities of the city. Plasticity in the brain, in the plastic object, and in the city, is both its radical potential and its vulnerability to or role in the catastrophic.

The urban theory with which this book is allied understands the city as more process than stasis, more an agglomeration of flows and exchanges than a circumscribed material environment, and more media ecology than a stable technological system. Plasticity is particularly good at capturing these ways of framing the urban without collapsing the very real embodied and physical phenomena of the city into an afterthought. It contains, as Malabou points out, the extremes of the concrete material form and the explosion of all form, which is both the most static and immobile and their opposites, an utter refusal, even destruction of the static.

As I also discussed in Chapter 3, plasticity offers a strong counter to neoliberal resilience discourse. Resilience discourse is particularly powerful in the city put at risk as the consequences of anthropogenic climate change, contemporary global capitalism, and rapid technological development make themselves increasingly present in the urban environment. Cities frequently employ "resilience" as a way to pitch new development projects, infrastructure expansion, or urban policies. Even financial instruments are said to be resilient: in 2018 Lloyd's of London and the UK's Centre for Global Disaster Protection proposed to bring to market four such instruments, including a "resilience impact bond" and a "resilience bond."[6] Architects and construction companies sell their wares under the banner of resilient design.[7] NGOs employ the language of resilience to justify their interventions and win grants (The Climate Resilience Fund, for example, and the Kresge Foundation's Environmental Program insist on "resilience" as an essential quality for facing the future and both have programs that target the urban world). Urban planners promise that their projects will increase resilience in the cities in which they work. In short, one does not have to look far to find someone proclaiming the virtues of, or promising to deliver the quality of resilience. As a neoliberal mechanism, resilience is framed as the necessary and desirable quality for people, for communities, for economies, and for cities. Resilience is an ideological project, beckoning people and communities to take on its form.

And it is not exclusively North American. Buenos Aires has a large-scale municipal public campaign, "Buenos Aires Resiliente," as do cities across the hemisphere:[8] Bogota, Santiago de Chile, and Mexico City have similar programs. It would seem that municipal governments are virtually mandated to parrot the language of resilience. As the impacts of climate change grow increasingly terrifying, resilience is over and over again the word in the mouths of city officials and the goal (often used loosely or in ways that can't be measured) of every major metropolitan area.

Kaiama Glover has described the ways that, in the case of the Black subject, resilience is bound not to reward but to repugnance. "For in the end," she writes, "resilience can only mean a capacity for suffering combined with a refusal to lie down, to stay down, and to die."[9] Glover looks at the figure of the zombie – a figure that emerges in speculative fiction, let's note – to think through the ways the poorer peoples of the Global South, and Haitians in particular, can be framed as resilient just as they are positioned as repulsive or repugnant, like the zombie hordes whose threat is existential.

None of this is to suggest that the policies, building strategies, or urban development that may materialize under claims of resilience are per se unsuccessful or even unnecessary. Rather, it is to point out that they are and will continue to be insufficient. It is not coincidental that part of the ecological disaster of the proliferation of single-use plastics lies precisely in just how resilient plastic is. A 2010 Ocean Conservancy report found that plastic bottles can take up to 450 years to decompose; fishing line, according to the report, has an estimated decomposition rate of 600 years.[10]

Cities require a more radical paradigm shift than resilience can offer, and as Glover, Malabou, Cooper and Walker, and others have pointed out, there are dire consequences for human bodies, for urban ecologies, and for much broader fields of life under the discursive regime of resilience. The discourse offers little room for structural critique, and too often resilience is employed as a means not by which to steel ourselves against what might be coming, but rather to perpetuate the belief that "we" won't be the ones to suffer, and worse, that some distant, darker others deserve the coming catastrophe, that some human lives aren't really human at all. Glover writes,

> The category of the human is jealously guarded and politically bound—and race is, of course, the social fiction most critical to maintaining this boundary. It is the last hold of visceral desire for difference enfleshed, visible difference that allows for the denial of relation and of proximity."[11]

Resilience discourse is one means by which that boundary is effectively and jealously guarded.

What is so central about Malabou's argument (and what is so urgently needed in contemporary radical practice in the urban Americas) is the capacity of plasticity to explode resilience in favor of a more responsible, resistant, and ethical frame. I return to it here because of what I see as the absolute necessity of plasticity in the urban environment. Malabou sites the possibilities afforded by plasticity: explosions of rage, refusal, and radical opposition to "a certain culture of docility, of amenity, of the effacement of all conflict even as we live in a state of permanent war."[12] Plasticity fights against a culture of flexibility (one that goes hand in hand with resilience). One of the vital contributions of Malabou's work, too, is that she insists that it is *our* brain, our plastic brain, that can and should change. This is a philosophy of plasticity that critically refuses to deny relation and proximity, mutual vulnerability, and precarity while nonetheless embracing antagonism and difference.

To insist that the city is not flexible, not resilient, but rather plastic also, following Malabou, opens up space for a resuscitation of positive destruction. If the catastrophic urban imaginations I have traced through case studies in this book are any indication, there is power in, at the very least, creatively envisioning catastrophe. Among its merits is its ability to destroy not communities, but ideological veils. To see plasticity in the city means to see in it "not only the creator and receiver of form but also an agency of disobedience to every constituted form, a refusal to submit to a mold."[13]

Communities in protest gather in city streets. They also form networks of support in crisis: this has been evidenced in almost every urban crisis the world has seen, although it is rarely covered in popular media accounts. Rebecca Solnit's *A Paradise Built in Hell*, for instance, traces the incredible ways in which communities of support, connection, and even joy have consistently arisen in the aftermath of major disasters. "Rising from the ruins," according to her work "is the ordinary human response to disaster."[14] The five disasters she studies all unfolded in urban terrain. In these moments, it seems, communities use their plastic brain to reach across divides, to rebuild together, to build common paradises in the bleakest of circumstances.

We might oppose the activist forms of imagining catastrophe in the urban world to the "creative destruction" first described by economist Joseph A. Schumpeter in his reading of Marx.[15] The process, internal to capitalism, according to Schumpter, perpetually destroys its old administrative forms and builds new ones. David Harvey has taken the theory up and pointed out the ways the process of urbanization toward the goal of managing surplus, and what he has called "accumulation by dispossession," have been among creative destruction's most essential aspects.[16] Creative destruction frames its own unfolding in the urban world and elsewhere as the perpetuation of economic development (though it never delineates the disasters, the real and terrible destruction, this development causes). Manuell Castells has also written about creative destruction, tying it specifically to the development of the megalopolis: This contradictory urban form is "globally connected and locally disconnected, physically and socially;" it has made entire swaths of the population redundant in the flows of the global informational economy.[17] Creative destruction has become a catastrophe of, if we follow Harvey and Castells, epic and global proportions. It remains, however, not just ubiquitous in popular conceptions of economy, but celebrated. As Alan Liu points out in his *Laws of Cool* "Schumpeter's dictum has become cliché (his other work and especially his prediction of the eventual demise of capitalism conveniently forgotten)" "in the thriving print and television journalism of business."[18] It is in creative destruction that we might see capitalism itself as plastic in its internal capacity for radical change. But, as Schumpeter knew, capitalism hides the true consequence of its plastic nature: that it contains within it at least the possibility of its own end, the explosion of its own form, not only of its forms of administration.

If we frame the city as plastic, there is room for ethical refusal that is not just the refusal to lie down, stay down, and to die, not only the refusal of one particular administrative economic structure in favor of another, but the refusal of ideological operations that turn an "us" against an other. There is the possibility, too, within a plastic city, to see alternatives to capitalist management strategies and the delirious logics of the debt form.[19] This is because plasticity is, as Malabou points out, a *common* materiality of our brain. A plastic city is only plastic in so far as it is common space, a space both formed by and through the common.

The power of plasticity is the power to explode the ideological frames that hide or refuse the common nature of the city, to demand the destruction not of livelihoods and lives, but of neoliberal policy in the city. It is a move toward subjects seen as both users of and builders of the city. That building and using are ongoing and perpetual. It preserves and respects the messiness so central to city life.

Plasticity also has the capacity to resist what Amin and Thrift have described as the tendency of a large number of urban thinkers beginning with Marx and moving through Simmel and Benjamin on through the work done today by the likes of Johnathan Crary and Paul Virilio, to view the city with nostalgia. These thinkers frame the city and its relationship to commodity capitalism as in the process of decay. Theirs is a particular form of catastrophic urban thinking. According to Amin and Thrift, in this kind of urban scholarship, the decay of the urban is linked to what they call four "blows" dealt to the city: money, thingification (that is, essentially, commodification), speed-up (the quickening pace of modern life), and mass media.[20] They propose a more tempered kind of urban theory, one that recognizes possibilities opened by the four of the blows: different forms of community formation, for example, engendered by new technologies, or different registers of exchange that exist within the money economy. They point, too, to scholarship that has revalued "things," since at least Bruno Latour's Actor Network Theory, and dispute what had been the assumption that "thingification" itself is necessarily linked to decline.[21]

The plastic city, however, is more radical than their proposal to reconsider scholarly approaches to the urban because it can do more than understand what the city is and does; it can demand its communal reshaping. Explosion: this is what the city needs if anything lovable about it is to survive. Explosion is necessary, in fact, to assure that the city can be built for and by its inhabitants. It lets those inhabitants be an "us" whose constitution does not depend on the refusal or subsumption of the other. Plasticity is our shared human condition; it is our shared vulnerability to suffering and our shared plight in the face of catastrophe, and it is also our shared power to make and remake the city.

Plasticity, unlike resilience, puts users of the city in the active position. Resilience is a quality, perhaps even a very desirable quality for a city, for a person, for a thing. Plasticity, as I argued in Chapter 3, is as much about *capacity* as it is about quality of being. Instead of simply responsive, the plastic city is active. If resilience is consistently framed as the quality necessary to prepare for disaster, plasticity is an ability to shape the world, to avert at least some catastrophes, to acknowledge the catastrophes that are and to refuse them, to stop them in their tracks.

Plasticity can be found in both Lefebvre and Harvey's "right to the city." It is evident in Thrift's defense of misanthropy as well. Haraway's *Chthuluscene* is plastic, and so, of course, is her cyborg. Guattari's ecosophy is certainly a plastic philosophy.

A legitimate, communal reframing of the city as plastic would mean not only a paradigmatic shift in the way we frame the city, and understand it, but also in the way we practice the city in our daily lives. Harvey's right to the city, that most valued and perhaps most neglected of human rights, is so important because it recognizes the common as the urban's true terrain. Wrested from the common, that right remains almost exclusively in the hands of very few: an elite class of urban developers and wealthy landowners who bulldoze their own policies into existence at the expense of the vast majority of city-dwellers. Mike Davis has pointed to the global consequences of the incredible wealth gap that has been widened by the refusal of this right to throngs of urban poor. Neoliberal restructuring since the 1970s has swelled the global slum population to massive proportions and increased and worsened the risks this class of urban dwellers are exposed to:

> The urban poor [...] are everywhere forced to settle on hazardous and otherwise unbuildable terrains–over-steep hillslopes, river banks and floodplains. Likewise they squat in the deadly shadows of refineries, chemical factories, toxic dumps, or in the margins of railroads and highways. Poverty, as a result, has 'constructed' an urban disaster problem of unprecedented frequency and scope.[22]

It is this ongoing disaster that plasticity, if fully realized as the essential quality of the city, could begin to address. If it is not, Davis points to a particularly bleak future: "Night after night, hornetlike helicopter gunships stalk enigmatic enemies in the narrow streets of the slum districts, pouring hellfire into shanties or fleeing cars. Every morning the slums reply with suicide bombers and eloquent explosions."[23]

The plastic city has the capacity for explosion, by all means. But, as Malabou points out, these need not be terrorist. They could, instead, be explosions of rage, of refusal – explosions that could reshape the city for the common, and restore the right to the city to its proper holders.

Notes

1 Roland Barthes, *Mythologies,* trans. Anette Levers, (New York, Hill and Wang, 1957), 195.
2 L. Lebreton et al., "Evidence That the Great Pacific Garbage Patch Is Rapidly Accumulating Plastic," *Scientific Reports* 8, no. 1 (March 22, 2018), 4666.
3 Dana Phillips, "Ecocriticism, Literary Theory, and the Truth of Ecology," *New Literary History,* Vol. 30, No. 3, Ecocriticism (Summer, 1999), 577-578.
4 Barthes, *Mythologies,* 193.
5 Catherine Malabou, *What Should We Do with Our Brain?* trans. Sebastian Rand. (New York: Fordham University Press: 2008), 5.
6 "Financial Instruments for Resilient Infrastructure" (Centre for Global Disaster Protection, October, 2018), https://forms2.rms.com/rs/729-DJX-565/images/Financial_Instruments_for_Resilient_Infrastructure_Technical_Report_October2018.pdf.
7 See, for example, the work of the national non-profit The Resilient Design Institute whose sponsors include roofing manufacturer GAF and the global architecture firm HKS: https://www.resilientdesign.org/about/

8 "Buenos Aires Resiliente | Buenos Aires Ciudad - Gobierno de La Ciudad Autónoma de Buenos Aires," accessed December 18, 2019, https://www.buenosaires.gob.ar/jefedegobierno/secretariageneral/buenos-aires-ciudad-resiliente.
9 Kaiama L. Glover, "'Flesh Like One's Own': Benign Denials of Legitimate Complaint," *Public Culture* 29, no. 2 (82) (May 1, 2017), 235–60.
10 "Trash Travels: From Our Hands to the Sea, Around the Globe, and Through Time" (Ocean Conservancy, 2010), https://oceanconservancy.org/wp-content/uploads/2017/04/2010-Ocean-Conservancy-ICC-Report.pdf.
11 Glover, "'Flesh Like One's Own'," 254.
12 Malabou, *What Should We Do*, 6.
13 Malabou, *What Should We Do*, 79.
14 Rebecca Solnit, *A Paradise Built in Hell: The Extraordinary Communities That Arise in Disaster* (New York: Penguin, 2009), 8.
15 Joseph A. Schumpter, *Capitalism, Socialism and Democracy* (London: Routledge, 1942), 82–83.
16 David Harvey, *The New Imperialism* (Oxford: Oxford University Press, 2003), chapter 4.
17 Manuel Castells, *The Rise of the Network Society* (Oxford: Blackwell Publishing, 2010), 436.
18 Alan Liu, *The Laws of Cool: Knowledge Work and the Culture of Information* (Chicago: University of Chicago Press, 2004), 2.
19 On capitalist delirium, see Melinda Cooper, *Life As Surplus: Biotechnology and Capitalism in the Neoliberal Era* (Seattle: University of Washington Press, 2008).
20 Amin and Thrift, *Cities, 32-33*.
21 Bruno Latour, *Reassembling the Social: An Introduction to Actor-Network-Theory* (Oxford: Oxford University Press, 2005).
22 Mike Davis, "Planet of Slums," *New Left Review* 26 April 2004, 5.
23 Mike Davis, *Planet of Slums* (New York: Verso, 2006), 206.

4 From the ground up

Aerial tactics and top-down strategies in Los Angeles

If you looked up into the smoggy skies of Los Angeles at just the right time in 2006, you might have been startled by a pigeon. Though the birds are commonplace in the city, this particular pigeon would have been an arresting sight because, unlike its avian peers, it would have been carrying a little backpack between its grey wings (Figure 4.1).[1] That bird, whether or not you had the luck to spot it, was part of a media project titled "Pigeonblog." The project was produced by the artist and activist Beatriz da Costa and a host of collaborators and participants. The tiny pack resting on its back was built to track the pigeon as it flew and to monitor pollution levels in the Southern California air. Users could hop onto the Pigeonblog website throughout the project's life to see where the birds were moving and what pollution data they helped to gather.

In its September 2010 issue, *The Atlantic* published an article detailing a different sort of animal tracking than the one playfully done by Pigeonblog: the contemporary practice of monitoring parolees and convicts using small devices attached to their bodies.[2] Thanks to the Global Positioning System (GPS) and Radio Frequency Identification (RFID) tools embedded in these devices, companies contracted by state and federal correctional institutions monitor the movement of those they often call their "clients" in real time, tracking them 24 hours a day. Some devices can even monitor the intake of drugs or alcohol by the people who wear them. This prison without walls, as *The Atlantic* article pointed out, is genealogically linked to the RFID tags used to track cattle in industrial farming since at least the 70s. These tags are used to deliver food to the animals when a signal is read by an electronic dispenser. They can gauge the movement of livestock, and even trace the animal after its death in the slaughterhouse as it moves through meatpacking plants and out to retailers and consumers.

Da Costa's work, like the carceral and commercial surveillance detailed in *The Atlantic*, is interested in tracking and mapping living bodies in their environments. The birds' custom-designed backpacks were equipped with "miniature air pollution sensing devices enabled to send the collected localized information to an online server without delay." (see Figure 4.2).[3]

> The pigeon "backpack" developed for this project consisted of a combined GPS (latitude, longitude, altitude) / GSM (cell phone tower communication)

Figure 4.1 A carrier pigeon with its GPS-equipped pollution sensing backpack
Source: Image courtesy of Robert Nideffer.

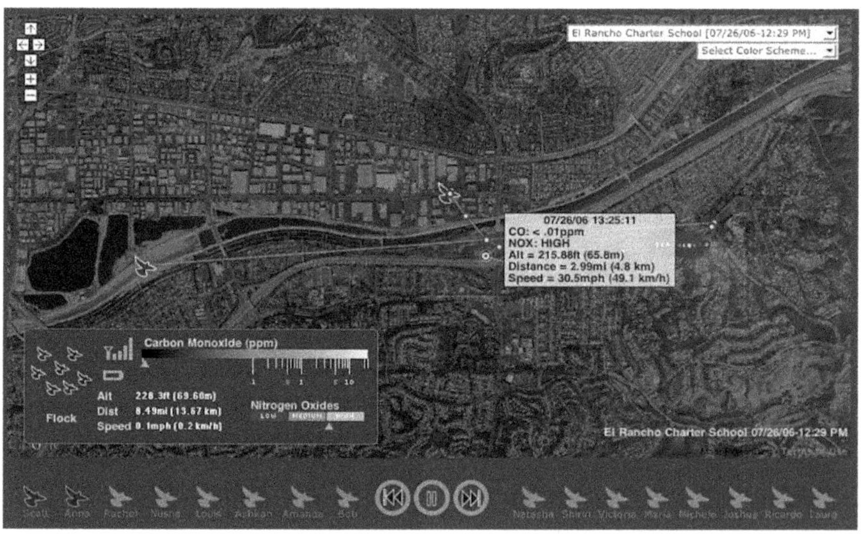

Figure 4.2 Tracking pigeons and pollution on the Pigeonblog site
Source: Image courtesy of Robert Nideffer.

unit and corresponding antennas, a dual automotive CO/NOx pollution sensor, a temperature sensor, a Subscriber Identity Module (SIM) card interface, a microcontroller and standard supporting electronic components. Designed in this manner, we essentially ended up developing an open-platform Short Message Service (SMS) enabled cell phone, ready to be rebuilt and repurposed by anyone who is interested in doing so.[4]

The tracking tools used by Pigeonblog, and those used for commercial and carceral surveillance, require a particular set of infrastructures to do their work. GPS depends upon a network of U.S. military owned-and-operated satellites that allow a device to be located by the triangulating signals. RFID technology depends upon a localized setup of radio antennas and, in the case of the cattle monitoring, various technologies that allow the feed mechanism to recognize a given tag and dispense food or withhold it accordingly. Both Pigeonblog and corporate firms like BI Incorporated and Texas Instruments that sell these devices also need mapping applications to locate cattle, "clients," or pigeons. They require Internet access as well as mobile communication infrastructures to move information about the location of those they track, to plot the routes and coordinate device maintenance. There is also necessary hardware that interfaces with these infrastructures. These technologies and infrastructures are layered throughout city spaces. They penetrate the vertical field, spread across urban sprawl, and pulse under the earth. They shape living bodies and guide movement.

The chapter that follows reflects on two distinct activations of popularly imagined catastrophes: Pigeonblog directs its efforts toward monitoring increasing pollution and the precarity (and potential toxicity) of the biological bodies that occupy and shape our urban ecologies. Governments and corporations involved in the surveillance of bodies, on the other hand, offer the (albeit false) promise of regulation and preemption of catastrophe: their narrative about the technology they use obfuscates the potential of revolt and disorder, even of waste, in favor of techno-capitalist efficiency. The efficiency companies and governments sell depends entirely on the perpetuation of an imagined future catastrophe that must always remain on the horizon but can never, in fact, arrive. Surveillance (the word I will use as a catchall here for commercial, actuarial, and carceral tracking of bodies) works toward biocapitalization, which offers the promise of preventing an imagined future catastrophe as the very mechanism by which to obscure the real catastrophes that its methods perpetuate: the extraction of capital from life on the ground, at the expense of both the bodies it tracks and the ecologies in which it acts. As detailed above, both the tracking done by Pigeonblog and commercial and state entities use similar technologies. Their employment of these technologies, however, works to contrasting ends vis a vis the future catastrophe they imagine, and they articulate the spaces they operate in differently. Pigeonblog posits an ongoing destructive process and records its unfolding. The state imagines a future catastrophe it promises to forestall. The real potency of da Costa's project lies in the way it subverts the very myths that surveillance needs in order to operate: perfect knowledge of a wholly readable city. As I will show, Pigeonblog converts the chaos that it finds on the other side of this fantasy – the messy, risky, and imprecise stuff of urban life – into an opportunity for more productive practical action against the power of the status quo, driven as it is by its desire to convert the life it manages into biocapital.

Pigeonblog and its counterexample in the work of surveillance produce competing temporal framing for the catastrophic. Pigeonblog, though obliquely, understands catastrophe as unfolding within bodies and ecologies as an ongoing processual phenomenon, not an event. That is, the catastrophe Pigeonblog positions itself within is the slow and structural catastrophe of the current biopolitical regime. Its violence is slow. Surveillance produces, instead, the foreclosure of catastrophe, and its foreclosure must be framed in perpetuity.[5] As we shall see, this discourse participates in the dominant paradigm of a total knowability of lively urban relationships, one that adheres to and perpetuates mythologies of precision, efficiency, legibility, accessibility, and freedom from risk even as it positions catastrophe as looming. It places technologies and infrastructure in the service of biocapital by maintaining an instrumental view of nature; it promises that available technologies and infrastructures can not only contrl and know life, but readily convert that life into surplus.

The concept of biocapital has been thoroughly developed by Kaushik Sunder Rajan in his *Biocapital: The Constitution of Postgenomic Life*.[6] In this work, Rajan traces the increasing intertwining of market forces and biotechnology. While I do not address the biotechnology industry or even the life sciences per se here, biocapital remains the term that most directly indicates the intertwining of regulatory forces and technologies acting on and through life under capitalism: at a variety of scales, life is made a site of the production of surplus value. Using technologies and infrastructures to regulate the bodies of cows, "clients," or pigeons, as both Pigeonblog and corporations and governments do, situates their endeavors well within the terrain of biotechnology, though perhaps biotechnology more broadly conceived than that Rajan studies.

Operating with very similar technologies and infrastructures and playing in similar urban ecologies, Pigeonblog complicates and resists exactly the mythologies of knowability and regulatability tied to biocapital. While surveillance insists on the infallibility of the technologies it uses, the exactitude of the locations it claims to pinpoint, and the total coverage of the infrastructures on which its work depends, Pigeonblog exposes the productive possibilities of risk and imprecision, of the fuzziness of borders and boundaries and the slippages and aporias that life poses to those who seek to know it. The project does this in order to advocate for and indicate the possibilities of a tactical intervention in contemporary structures of power within the urban environment. While this chapter's arguments are specific to Pigeonblog, all of what follows is bracketed within a larger examination of the value of an aerial perspective on the urban.

Los Angeles is consistently discussed as a city that grew out instead of up – urban sprawl par excellence. Let me remind readers that it was Los Angeles County, seen from above, that inspired sociologist William Whyte to coin the phrase "urban sprawl" in 1958.[7] As Mike Davis points out, the city's ferocious outward growth and development even became the source of national scandal.[8] The discourse surrounding Los Angeles and sprawl usually focuses on the outward spread of the city from the ground level. Travel through the city is nearly always horizontal and over ground. The commuter flights between airports in larger Los Angeles, Orange, and Ventura counties never materialized quite as critic Reyner Banham imagined they would in his Los Angeles. But there is, of course, a view from above in Los Angeles, and access to the vertical can offer a critical perspective on the mediation of urban ecologies. Altering the critical position on the landscape of the city (whether via pigeons or Google's aerial views) decenters the atomized perspective of the lone driver or pedestrian to include actors and agents in the sky above. Positioning ourselves in the air above the city, though, also means being alongside the city planner and the military satellite. As will become clear in the objects this chapter studies, the tools and strategies of capital are also often adopted by the urban activist who reamakes them as tactics of subversion. But in and of itself, an aerial perspective on the city is no guarantee of subversive value. Instead, it should be thought of as a means by which to leverage some of the language of a biopolitical approach to the urban against itself. It is to insist on the capacity of signs to work directly against themselves to produce opposite and alternative meanings: Pigeonblog, employing a surveilling gaze from above, actually operates against the logics of biocapital that guides the (also aerial) view of surveillance.

The chapter is divided into three sections, each devoted to umbrella concepts that Pigeonblog plays with as part of its "pursuit of resistant action."[9] These are infrastructure and technology, pathology, and ecology. The three interplaying concepts are central, as well, to the material effects of the epistemological milieu in which both surveillance and Pigeonblog are located. The first section examines the ways technological communication and transportation infrastructures both help to build the ecology in which the project is embedded and offer an opportunity for Pigeonblog to actively demonstrate alternatives to dominant mythologies within that ecology. Pigeonblog is interested in new means of producing knowledge toward resistant ends, not in mandating those ends or producing them. It is in this way that the project persistently resists the kind of "firmative" speculation employed by surveillance in favor of the affirmative speculation of playful urban intervention.[10]

The second section looks at how Pigeonblog activates the pathological. Pathology is, here, a way of thinking through the relationship between popular notions of "otherness" and biological precarity or biological catastrophe for human and nonhuman animal bodies. The project plays with pathology in ways that run directly counter to the instrumental control of both biological life and complex ecologies: it is in the pathological that Pigeonblog situates itself most obviously as work made within and through catastrophic thinking.

The third and final section, on ecology, elaborates the concept both in terms of the layered relationships between new media technologies, living bodies, and the earth, and also further develops it as a functional critical model that can describe the way contemporary power functions, particularly as exercised over the relationships between bodies and technologies.

What is particularly valuable in using ecology itself as a critical model is its insistence on nonhuman actors – plants, animals, technological infrastructures, architectures, and more – as integral to the making not only of urban worlds, but of political lives. The relationships between subjects and each other, between people and the ecologies they inhabit and help to make, demand a critical language that can accommodate the force of these actors, can adeptly engage with them, and can do so without re-centering humans as somehow separate and solitary subjects. Pigeonblog operates ecologically. This is tricky work, but necessary to produce a faithful outline of what is at stake in contemporary cities, and what paths of action might be open. What I hope to offer, then, is an examination of two aerial approaches to urban ecologies: one that aims to open up the air as space of critical participation and invert the instrumental approach to life required by biocapital, the other that wants to use it, rather, as a mechanism of power over, and extraction of capital from, life. In one catastrophe is also an opportunity. In the other, looming catastrophe is an opportunity for profit but it must be, via the status quo, forever forestalled.

Infrastructures: Mobility and modulation/fixity and precision

Pigeonblog and surveillance are situated within the complicated relationship between the global and the local, itself charged by competing discourses about the reach of global infrastructures. Firms that produce and sell hardware and software for surveillance, as well as the U.S. military and others, posit GPS as a technology that allows for the total revelation of spaces, places, and things. As the U.S. government describes it on a website devoted to the subject, GPS is "a global public service brought to you by the U.S. government," that "provides users with positioning, navigation, and timing (PNT) services."[11] GPS has both military and civilian users, but according to gps.gov, "the civilian service is freely available to all users on a continuous, worldwide basis."[12] "Continuous" and "worldwide" suggests the total visibility of life on the ground. For the offenders carceral surveillance hopes to track, such global coverage, at least in theory, leaves them no hiding place. Their movements are always and everywhere monitored: what catastrophes they might cause are, in this view, preempted.

The discourse of total, global vision flattens the local to mere coordinates of longitude and latitude. What matters for surveillance is not communities or regions but easily plot-able zones of movement and boundaries. BI Inc., one major seller of surveillance hardware and software, explains that its "ExacuTrack" ankle bracelet captures "data points [...] throughout the day" that BI then compares "against the client's 'boundaries' and 'schedules' when he or she returns home. For many corrections agencies, this is an effective, time-saving solution that requires less manpower but still provides precise tracking details."[13] Note the frequency in BI's language of relatively abstract terms for the space in which BI's clients are tracked, coupled with the simultaneous insistence on absolute precision and efficiency. In effacing the specifics of the local, BI can also efface the challenges the local poses to the mythology of total visibility that the company uses to sell its products and that biocapital itself employs. The result is the production of a picture of tracking technology that converts convicts into clients and in which the prison dissolves from a specific location and architecture into a diffuse "everywhere." Exact locations and precise movements are presented as wholly knowable in this view, and they must be such because it is this supposed knowability that converts lively bodies into capital through promises of containment.

Pigeonblog, on the other hand, uses GPS to counter this mythology in a number of ways, aiming for at least partial visibility, certainly, but site-specific, local visibility that engages the complexity of actual areas rather than ignoring or covering over location-specific information. While the project harnesses supposedly global infrastructures and might aim for a global audience it never lets go of the richness of the local. Pigeonblog's tight bond to the local is maintained even when its richness poses problems for the project – pigeons may fly out of general bounds, for example, or into areas in which accurate GPS data cannot be gathered. This means that catastrophe (in its case, the catastrophe of toxic pollution), as it is explored by Pigeonblog, has distinctly local implications: where it unfolds matters.

The specific areas where Pigeonblog sought to monitor pollution are often neglected by governmental regulating bodies, which have considerable trouble getting accurate levels at low altitudes and which avoid pollution "hot spots," such as areas directly neighboring highways or power plants. These areas are, as Pigeonblog points out, very often low-income and minority-dominated neighborhoods, and thus their neglect by government regulatory agencies disproportionately places the burden of pollution-related health problems and the work of monitoring them on already disadvantaged sectors of the population, with the fewest means and legal recourse (particularly in the case of noncitizens).[14] Pigeonblog thus marks local spaces that the South Coast Air Quality Management District is blind to. It outlines animals, plants, and so forth on the ground and in the air, in particular neighborhoods with particular textures, using GPS technology. The blind spots it unearths are politically charged spaces – ignored for reasons that are connected but not limited to the reach of infrastructure or GPS locatability. This work is vitally important in Los Angeles. As of this writing, the city is currently ranked first by the American Lung Association in ozone pollution, fifth in year-round particle pollution, and seventh in short-term particle pollution.[15] And as Pigeonblog works to underline, this toxicity does not affect the various life that is rooted in L.A. equally. Where the ExacuTrack hopes to efface specificity and distinct location (paradoxically through the claimed exactitude of its technologies) titling all those it tracks under the politically neutral and obfuscating title "client," Pigeonblog hopes to unearth exactly the ways in which particular spaces are ignored, invisible, politically charged, or biologically toxic. The project does use GPS to see something that was previously hidden, but unlike surveillance, its vision is not total or totalizing. It highlights the ways in which the material particularities of these areas create problems for certain kinds of knowledge, specifically knowledge about pollution levels. GPS can help give the project access to these areas, but the point is that GPS alone does not render the world whole or wholly visible. Pigeonblog's resistant capacities are tied to its refusal of perfect visibility in favor of a more dynamic, and more playful, production of knowledge on differing scales.

Because access to accurate GPS readings is always at least partially under the control of the state (the U.S. Air Force operates the satellites), using the technology is always partly a political place-making. This is easily seen in its strategic uses for military conflicts. GPS helps the state track bodies on the ground, plot bombs, and execute surveillance – in short, it marks the place of the enemy by marking life on the ground, by outlining boundaries, by locating in space and time the life it seeks to regulate or exterminate. (And keep in mind that this tracking is done almost exclusively "from above"). Lisa Parks has described "vertical hegemony, [...] the ongoing struggle for dominance or control over the *vertical field*, which here includes combinations of terrestrial, aerial, spectral, and/or orbital domains."[16] Looking specifically at developments in U.S. policy and practice after the 2001 attacks on the World Trade Center in Manhattan, she argues that the vertical is an essential site for the negotiation of power. She writes:

> The struggle for vertical hegemony is undergirded by the assumption that controlling the air, spectrum, and orbit is tantamount to controlling life on earth. Vertical hegemony cannot simply be achieved "out there" in open skies. To register or take effect, it must be communicated though and materialized as part of culture on earth.[17]

Surveillance tracks bodies as a mechanism to maintain vertical hegemony – it employs material in the vertical field. It regulates the life of those it tracks, preventing movement in certain areas. It monitors and controls its clients' bodies for an express biopolitical end. Pigeonblog, however, is not looking to regulate the bodies it tracks or the spaces in which those bodies move, nor is it primarily an attack on the environmental pollutants that GPS helps the project monitor. The project, rather, uses GPS to outline possibilities for "resistant action" against a discursive regime that employs mythologies of precision and freedom from risk to perpetuate the asymmetrical exercise of power. Where surveillance wants to use GPS to produce an always seen, regulated, legible world, Pigeonblog unearths blindspots in the logic of such technologically mediated surveillance. It maps the power structures hidden by notions of universal access are instead mapped by Pigeonblog. Where GPS "promises," in the worlds of Architect and media critic Karen Kurgan

> that people and their vehicles will never get lost; missiles and bombs, as well as airplanes, will land exactly where they ought to; and a world of stationary objects from telephone poles to wetlands to private homes, will be fixed, once and for all in their proper places,"[18]

Pigeonblog sees another, far less precisely locatable world in which the use of GPS is socially and politically dynamic. Kurgan's tone here is tongue in cheek, and her work traces some of the ways in which GPS cannot, in fact, precisely map spaces, particularly in areas with very tall buildings or in bad weather. Pigeonblog uses the technology not to create a stable topology, but to suggest that sometimes avoided or ignored places require critical attention, that the scientific knowledge that plots data and makes averages may miss or erase knowledge of the spaces in between. What can be located as a specific site via longitude and latitude is itself already mobile, already moving in the project. GPS cannot stabilize the structures in which it works. It does not render spaces wholly legible in part because it reduces them to triangulated points on a global grid: there is no real texture there. With this in mind, the "place" that Pigeonblog helps to form is itself considerably more slippery and less visible than either the U.S. military or BI Incorporated might admit. In its very use of GPS, a technology that surveillance uses to claim specificity and precision, Pigeonblog raises questions about imprecise data and fuzzy borders, thus helping to dislodge the fictions of legibility and stability on which the extraction of biocapital under the current regime depends.

In fact, GPS as a technology has never really fulfilled its promise of total accuracy and total precision, particularly to civilian users. Until May of 2000, the U.S. Department of Defense imposed a policy of "selective availability."[19] Prior to this date, the transmissions from the U.S. satellites that make GPS use possible carried with them intentional and unpredictable sources of error. This practice of "intentional degrading of civilian GPS" was stopped and replaced with region-specific, "security-motivated" degrading.[20] Clearly, locatability and total visibility are far from universal even if they have been more attainable since the policy change. BI might use GPS and so might da Costa and her collaborators, but the technology is not available to all. It is easy to imagine, for example, the perils a project like Pigeonblog might face if it were executed in Syria or China. There are also problems that arise when using GPS in certain types of terrain. A study conducted after selective availability was abandoned found that some types of terrain, particularly mountainous areas and those with dense canopy coverage, interfered with the accuracy of GPS.[21] This fact is somewhat obscured in the materials produced by the likes of BI Inc. They claim to unearth potential hiding places by using a "mix of technologies."[22] Pigeonblog does not direct its users to technological failures directly, but the larger project indeed acknowledges and is interested in error and inaccuracy.

In an interview I conducted with her, da Costa was clear that the project was a "proof of concept," rather than an effort to compile any kind of comprehensive data set.[23] She was also quick to say that the data she collected may be inaccurate or unreliable. But the real indication of the difference between surveillance and Pigeonblog is in the two distinct activations of their media ecologies. Start with their names: the BI monitoring device, "ExacuTrack," has little in common with "Pigeonblog." The former is meant to evoke total stability and accuracy, the latter a much more mobile engagement with users and birds. Pigeonblog is an exploratory use of GPS, a speculative project interested more in the investigation itself (the blog format certainly indicates an informal and exploratory kind of critical reflection) than the accuracy or specificity of the results. The project mission statement, packed with questions that the work itself makes no direct attempt to answer, is another marker of the project's investment in speculation and exploration rather than fixed or total knowledge production. Even in its political position, Pigeonblog is invested in possibility and potential more than hard data or results. It is a project "in pursuit of resistant action," a project in progress and thus unfinished and unstable. This echoes, as I will discuss later, the way the work frames catastrophe: that it too, is unfolding and processual. Situated in the urban milieu, Pigeonblog also indicates an understanding of the city itself as a more of a mobile and processual constellation of phenomena than an ordered, static system.

Pigeonblog's version of resistance is tied to the bird tracking it does in neglected aerial spaces. Philip Agre's work is useful here: Aligned with the Deleuzian model of contemporary power, Agre suggests that emerging regimes of control have moved away from the panoptic spatial metaphor of surveillance and toward networked "capture" of data. This move, which Agre links to the "historical dynamics of a market economy," allows surveillance to move horizontally in decentralized and distributed networks (a rhizomatic, rather than arboreal command structure).[24] Power collects data almost everywhere. Pigeonblog is situated comfortably within this model. That is to say that the project, because it is invested in the collection of data in terrain that is ignored by environmental regulators, is invested in distributed data capture as, at least sometimes, a net positive for the production of counter-narratives to biocapitalism. The surveilling eye is active in the project, even if it belongs to the birds, or the cloud of users following the birds via the blog. This is perhaps the darker side of the work's speculation and of many contemporary models of urban intervention. It is here, too, that Pigeonblog displays a certain amount of playful pessimism. Piggybacking on the infrastructures and usurping the vantage points available to power means, at least partially, ceding to them. I do not intend to suggest that this negates the critical intervention of the Pigeonblog work. Quite the contrary: part of the project of this book is to insist that certain kinds of urban speculation on catastrophe can and do come in packaging similar to those used by the ruling regimes. This can actually make them more productive and resistant political acts, buoyed by the mirroring that they do of the structures at which they also take aim. Pigeonblog is by no means anti-capture, or even anti-surveillance per se, but it employs these phenomena against their popular construction: to unveil rather than mystify, to make messes rather than to produce order.

Along with GPS, da Costa's project also uses SMS technology to move the data the pigeons and their packs collect to on-the-ground collaborators. The dominant discourse surrounding this technology and its infrastructure, like that of GPS, boasts of universal, global connectivity. SMS, though, unlike GPS, is not a technology so directly controlled by the state, but rather by a combination of state institutions and private corporations. Sending information via wireless networks means, as Ted Kane and Rick Miller point out in their discussion of telecommunication in Los Angeles, is thus tied to corporate providers as well as governmental institutions regulating telecommunication:

> Like the freeways before them, wireless networks hold out the promise to spatially liberate the citizen by connecting the city without undermining the autonomy of the individual citizen. The wireless network spans the extensive geographic breadth of the city, creating new connections where the freeway left off. However, this new infrastructure is no longer a part of a regional plan; rather it is carved out by private corporations competing for market share. These are commercial enterprises, which by their very nature are competitive yet redundant. Today in Los Angeles four telecommunications corporations – Verizon, Sprint/NEXTEL, T-Mobile, and AT&T – operate simultaneously, stitching together independent wireless phone networks based on their own strategic models.[25]

So while Pigeonblog works against the claims of total visibility, total efficiency, and precision of GPS, it also navigates global corporate giants who promote claims of global connectivity and instantaneous access. In 2013, a Verizon ad campaign put billboards up across Los Angeles with the tag line "Rule the Air."[26] TMobile, selling itself as "America's Largest 4G Network" added its own slogan during this time: "Share life – in an instant."[27] These advertisements, and many like them, are meant to suggest an instantaneity of connection in a boundless everywhere in which any user with the right provider can connect, and of course, they together suggest that it is not just governments in pursuit of vertical hegemony, but corporations as well.

There is something particularly charged in this when we consider Los Angeles as an urban environment. Perhaps the first truly postmodern city, L.A. became so in part because of its transportation and communication infrastructures. Freeway and real estate development have formed it into a sprawling expanse of neighborhoods linked together by lines of traffic that, along with the enormous thoroughfares and endlessly shifting flows of gentrification, also isolate them. In many ways, L.A. itself is an agglomeration of partially functional networks that simultaneously mobilize individual actors who are also linked, isolated and blocked by competing infrastructures. The freeways link sprawl together, allowing for travel between neighborhoods, but also create huge barriers to pedestrian traffic that might have or once did bring neighboring communities together.[28] The heavy reliance on automobiles in the city also means that as people travel through the city, they are often isolated from each other, moving through space in their own vehicles. Urban life is thus funneled through channels, parceled into distinct locations and (as municipal governors, corporations, and others assure us) thus secured.

Norman Klein writes specifically about the relationship between freeways and other development in 20th-century Los Angeles motivated by what amounts to white fears of their nonwhite neighbors:

> Much of that anxiety centered, at first, around fears of Mexican neighborhoods close in then, later, about African Americans. Soon it became an architectural necessity. Cars moved along guarded freeways. Civic buildings needed security against too many winding alleys or hidden public places.[29]

Urban infrastructures, in many cases, were what both liberated the citizen in the city, allowing her to move quickly and easily around, and what secured her. But these same infrastructures could open up avenues through which perceived threats might travel. They could provide security, or be its breach.

The discourse on mobile telecommunication technologies and infrastructures follows this same logic: it is said to be both liberating and potentially the site of vulnerability to or production of certain risks. Total connectivity lets everyone roam free and remain linked into the network (that connectivity, after all, is in the air all around the city). For surveillance, this means its clients remain clients in view, unable to slip into an alleyway or a dead-zone to once again become criminal. For Pigeonblog, though, global connectivity does not mean increasing security. First, because the pollutants it seeks to track are also linked to the infrastructures it uses. The manufacturing of cell phones, the construction of towers throughout the city, the transport of goods along the highways and rails all produce pollutants. The infrastructures that keep the city and its inhabitants connected can themselves prove biologically dangerous. There exists a close relationship between biological precarity and urban infrastructures. Pigeonblog marked this by declaring that part of its work was to monitor pollution where local regulators would not: next to power plants, refineries, and highways, which are exactly the locations directly responsible for the maintenance and production of communication and transportation infrastructures. The possibilities of connectivity that the project relies on are also partially the cause of the problems it seeks to address. Sharing life "in an an instant" produces pollutants that may become a threat to life. "Sharing life" also indicates, despite itself, a certain shared vulnerability to the toxic. Pigeonblog points to risk. Its project is directly tied to tracking it. But this tracking is done in the service of outlining, not containing or obfuscating its presence in the city.

Let me add an anecdote, as well, as a means of exposing some of the aporias in the equation between total connectivity and security that surveillance might want to cling to, and that illustrates the close relationship between biological life and telecommunication systems that put certain inhabitants of the city at risk. Pigeonblog ran into some interference in the project from the cell carrier it used. In our interview, da Costa playfully described her interaction: "Phone companies think you're 'up to something' when buying so many sim cards at once under one name... Beatriz Bank Robbery or Da Costa Terrorist Cell."[30] While da Costa's tone here was jocular, the project's fray with its carrier is one more indication of the penetration of biopcapital into the regulation of access to technologies of communication. Where urban planners sought to minimize the perceived threats of minority communities and hidden spaces in Los Angeles, here telecom giants seek to limit the perceived threat of terrorism. Both are aimed at regulating and securing life.

Matthew Fuller, commenting on David Lyon's concept of "actuarial surveillance," writes:

> Lyon's insight in naming actuarial surveillance is substantial in that it makes the systems of contemporary control as processes of modulation visible at the scale both of everyday life and of systems operating on and in life at a more macroscopic scale and intensively and extensively monitored over time. Think of life insurance or of epidemiology as models, riding and shaping waves of life and death via statistics and millions of interlocking probabilities.[31]

Fuller's reading of Lyon here indicates two important components in the contemporary exercise of power. Modulations in everyday life and the way it is lived, who is communicated to and how, who controls, regulates, or prevents that communication, where a body can or cannot travel and how it manages to get there or how it is prevented from doing so are perhaps easier to see as means of control. Systems "operating on and in life," though, are also part of what power uses to sustain itself. The kinds of risk assessment a cell carrier makes when it agrees or refuses to sell a stack of SIM cards to a customer is an indication of the transformation of capital into biocapital and is materially experienced as the access or the denial of access by potential consumers.

It's worth echoing here as well that the "millions of interlocking probabilities" relied on by insurance agencies or healthcare corporations also partake in the logics, discussed in Chapter 3, of climate science. And, as with that "'conjectural science' that uses mathematical models in the hope of rendering the fluidity and hypercomplexity"[32] of vast interlocking systems legible, as Horn describes it, what BI, or TMobile, or Pigeonblog come up against is an epistemological problem posed by enormous and complex ecological, urban, and social systems. What distinguishes Pigeonblog is its assertion first, that the working models of monitoring pollution in the city are failing to track or address the complexity on the ground, and second, that its own project is necessarily partial. Pigeonblog acknowledges complexity, even acknowledges that urban terrain may not be fully knowable. Instead, it offers a way of playing in the city that situates itself in local participation in the urban ecology.

The pollution Pigeonblog seeks to monitor also indicates that everyday life and biological life are not separable categories. Information and exchange are pulled from their virtual ether back into the messy stuff of life. This exposes the logic of a particular biopolitical regime: total spatial liberation for citizens of the technofuture is the flip side of the (real or imaginary) covert terrorist cell. Life is free to roam. Life is at risk. To return to BI Inc. tracking the condemned bodies of their clients or the bovine bodies in the food supply, it is easy enough to see that the technological regulation of biological life is supercharged by politics. For Pigeonblog, the trick is to expose the interlocking ecological systems at work in order to offer a possible shift in the production of knowledge about those systems. For BI, the trick is to cover over the avenues of resistance afforded by the very technologies the company uses to extract profit from life, to refuse error and aporia in favor of a false but profitable and persuasive discourse that claims precision, freedom from risk, accuracy, and efficiency. Both depend on a bird's-eye view of the city: a place in the vertical field. In the case of BI's client tracking, that view is meant to give to police the kind of perspective afforded to the guard in the central tower of Bentham's panopticon. This is the position that is to offer the most power to the policing mechanism, such that in the end the constant and total visibility (from above) turns the prisoner's own gaze against himself. The old Foucauldian model still lingers. For Pigeonblog, the bird's eye view offers access to knowledge of something in the air that otherwise remains unseen. It's the (literal) bird's-eye view that helps the project look up, perhaps even look back at power. As discussed previously, too, the way in which Pigeonblog lingers in the camp of capture makes it a project that understands both the old (though it seems, not yet gone) Foucauldian model of surveillance is operational alongside and with the capture model. It remains to be seen if the work Pigeonblog initiated will help bring into being a city that distributes the power to see or remain hidden radically enough (in a multiplicity and diversity of users) to undermine biopolitical control. But the project does clearly outline and playfully upend the models of surveillance and capture that undergird contemporary biopolitics.

One gloomy speculative leap to take is to imagine the increased use of drone aircraft in urban policing as akin to the work Pigeonblog does. I am not here interested in clearly answering the lingering questions posed by the problem of the aerial view. By juxtaposing Pigeonblog with carceral and actuarial surveillance, though, I do want to indicate that the contemporary world demands that urban political action start thinking about just what it can do with the urban views to which it has access, and about who might already be taking a look. Parks' work on vertical media makes this more than clear at the global scale, but it is equally true at the local: vertical space has very material consequences in the city, on the ground.

Pathology: Alterity, toxicity, precarity

If there is a dark speculation on surveillance embedded in the work of Pigeonblog, there is also a speculation on the pathological. As a starting point, Nicole Shukin's discussion of "biomobility" is useful. Shukin develops the concept as the flip side of telemobility – the supposedly effortless and instantaneous ability to transmit information, affect, even selves through communications networks. Biomobility is the threat that accompanies telemobility, the possibility of disease leaping across spatial and species borders via some of the very same infrastructures:

> If telemobility traffics in the promise of a 'painless transmission' of affect through seemingly ethereal global networks, with biomobility the substance of virtual communication reappears in the pandemic potential of communicable disease. Biomobility names, in other words, the threat of telecommunications' pathological double, the potential of infectious disease to rapidly travel through the social flesh of a globally connected life world.[33]

If both the devices used to track prisoners and parolees and the pigeon pack function in the discursive space of telemobility, the animal bodies they track certainly tug them into the "social flesh" of the world. Both Pigeonblog and surveillance are interested in latching on to exactly this social flesh. The distinction to be made between them is how they construct a pathological body. Surveillance wants to promise, even as it harnesses "ethereal global networks," a kind of instrumental control of the body. This is part what vertical hegemony pretends to do. Pigeonblog, conversely, wants to excavate the potentially troubling and potentially liberating relationship between global infrastructures and the fleshy stuff of life. By laboring with and through animal bodies in an effort to track the sometimes catastrophic material impacts (in the form of pollution, for example) of infrastructural development and use, the project outlines both the ways that life can offer paths to resistance through these infrastructures, but also can be placed at risk by them.

The analysis of the pathological presented here is geared not toward the concept's medical construction, but to its mobile meanings within the social world in which Pigeonblog plays. Contingent on institutional desires and shared cultural fears on the one hand, and active and unpredictable biological agents on the other, the pathological is an unstable term. Its use here is twofold. The pathological functions, first, as a way to describe what are imagined as precarious relationships between animal and other bodies: the fear of disease that could cross from pigeon to human, that pollutants could cause illness in populations situated in the "hot-zones" that Pigeonblog tracks. The pathological works in the construction of neighborhoods and populations as themselves pathological, toxic and potentially threatening to other (typically white, wealthier) groups. It is also used here to describe the construction of otherness that so-often accompanies both social relationships between human actors and those across-species boundaries. Otherness is often figured in terms of desirability but also of pathology in both of these fields.[34] Pigeonblog takes a distinct political stance on this matter, asking that the pathological enter the field of political discourse not as the periphery that constitutes the center or the other that constitutes the self, not as the abnormal that constitutes the normal (although all of these are ways the pathological is constructed and functions). Instead, Pigeonblog wants to outline the pathological as both active and uneasy. "Pathology" can facilitate mutation in itself and the ecologies within which it is built and can lead as readily to productive and alternative participation in the making of knowledge and political discourse as it can to destructive and dangerous ends.

Two discursive nodes prove useful to examine in terms of the pathological in the following section: the pathological animal (in our case, the pigeon) and pathological space (the "hotspots" of environmental toxicity that were among the Southern California locals monitored by the project). Let us begin with the pathological animal. Pigeonblog positions the birds with which it works as "co-producers" and, in the coordinators' writing about the project, they have a very particular hope for the kinds of questions such coproduction could begin to answer:

> When thinking of pigeons, people tend to think of the many species found in urban environments. Often referred to as "flying rats," these birds and their impressive ability to adapt to urban landscapes isn't always seen in a favorable light by their human co-habitants. At least by association then, Pigeonblog attempted to start a discussion about possible new forms of cohabitation in our changing urban ecologies and made visible an already existing world of human-pigeon interaction. At a time where species boundaries are being actively reconstructed on the molecular level, a re-investigation of human to non-human animal relationships is necessary.[35]

In an express effort to offer a figuration of the pathological animal as something with positive political potential, Pigeonblog responds tactically to the reigning discursive regime. Instead of the pigeon as a "flying rat," a carrier of disease, or a public nuisance, the project sees in its coproducing birds a recuperative potential. By collaborating with these particular animals to monitor pollution, Pigeonblog uses a pathologically charged figure to mark and reveal biological risks. Pigeons, often seen as carriers of disease, become refigured as participants in the fight against disease-causing agents. As counterexample, Texas Instruments' specific cattle-tracking technology (or rather, the discourse around it) positions the threat of the pathological animal as the thing it can isolate, control, and dispatch. It encourages ranchers to use the technology to help ward off pathology because the threat it poses to animal life threatens the profit that comes from converting that life into capital:

> Texas Instruments' RFID technology helps ranchers round up stock more efficiently, provide feed and water at optimal locations when necessary, and

> even handle some basic health monitoring such as the frequency with which animals visit feeding stations. A decrease in frequency can be an indication of illness. When livestock is consistently well-fed and easy to locate, the value of each animal is maximized both in terms of market timing and the data that accompanies the carcass. RFID technology is frequently deployed at automated feeding stations and in slaughterhouses so that the animal's carcass can be accounted for well down the segments of the retail chain. Import/export considerations can also be monitored to assure customs authorities that the animal did not suffer from a disease.[36]

The control that RFID technology promises its users over animal life is directly linked to the data the ID can tie into the animal body. Ranchers can, Texas Instruments claims, use technology to halt pathology, to maximize efficiency and profit, even to make the vast routes of commerce that knit together the global economy and the global ecology safer. Technology is offered as a way to completely contain pathological risk. Throughout this discourse, the threat of sickness that could come from some relationships to these nonhuman animals is distinctly present. The technology Texas Instruments wants to sell, according to the corporation, can make these relationships safe and (for the biocapitalist whose interest is selling cattle) productive.

In highlighting a very different sort of productive potential in human/nonhuman animal relationships, Pigeonblog works against constructions of those relationships as inherently pathological without positioning the pigeon as an idealized figure of human–animal kinship. The project seeks expressly to "develop mutually positive work and play practices between situated human beings and other animals in technoscientific worlds."[37] It envisions human–animal kinship as both potentially productive and potentially pleasurable.

Shukin writes that "the fetishization and pathologization of interspecies kinship are twin expressions of power that do violence to existing and future collectivities of humans and animals."[38] Pigeonblog's coproducing birds are neither typical domestic pets nor the exotic creatures of a liberal fantasy of animal/human kinship that Shukin discusses. They are urban animals whose ubiquity in their local ecology has made them either the subject of urban fears (keep in mind that the project was situated well within a media environment rife with the terrors of Avian flu, SARS and H1N1) or nearly invisible to their human cohabitants. It is worth pointing out as well that two of the stated goals of the project are to bring to light an already existing relationship between pigeons and humans and to offer new political alliances between the birds and their human handlers. Da Costa and her collaborators write that "at least by association [...] Pigeonblog attempted to start a discussion about possible new forms of cohabitation in our changing urban ecologies and made visible an already existing world of human-pigeon interaction."[39] This is typical of the project's response to dominant paradigms of the pathological animal. Da Costa sought out ways that birds and humans were and are kindred, were already in relationship, albeit one that is hidden or ignored. The point is not a fantasy of human and nonhuman collaboration, but an excavation of the possibilities already built into our urban media ecologies. It is an advocacy for a new means of producing knowledge about our worlds by engaging these ecologies; it allows for playful participation instead of instrumental control. Where Texas Instruments tracks cows to maximize profit and efficiency, to limit risk as it claims to control biological life, Pigeonblog tracks its pigeons for purposes of play and production it hopes will benefit both the birds and their human collaborators.

In the project's practice, the birds did pose major concerns for their human counterparts. In my interview with da Costa she explained some of the challenges:

> It was difficult for me to work out my personal ethics here since obviously, no matter how you turn it, I am higher up on the hierarchy. I can theoretically digest a pigeon, not the other way round, so, seeing how to interact, how to sense if they were 'up for it' or not was a huge part of the project and [...] one of the main questions at the beginning was how humans can work with animals for whatever 'cause' is important to them. "How can we team up?" Not for agriculture, or transport, or sport or anything, but for cultural activities [that] subscribe to a certain political commitment.[40]

The inability of human project coordinators to know their nonhuman counterparts or to gauge desire, suffering, or pleasure in them was cause for da Costa's concern, particularly given the clear imbalance of power between the pigeon and human participants. In this way the pigeons remained, in some ways, the "other" to the human coordinators of the project. Working with the pathological animal, even in order to invert its construction as pathological, cannot erase the moment of facing the otherness of human and nonhuman creatures. Clearly, though, the project attempts to respond ethically to the call the other makes when we face her. This is, perhaps, the most radical trajectory Pigeonblog points us toward – co-production and collaboration across species. As a gesture, it might be described as utopian; even as it plays with catastrophe, Pigeonblog speculates on a collaborative, not post-human, but post-human-only urban world.

Pigeonblog refuses to speak for the birds, but instead tries to explicitly mark their articulating power in the space of an urban media ecology along with their vulnerability within that ecology. As Shukin writes, reading both Judith Butler and Jacques Derrida, "the zoosphere emerges out of a shared ability to suffer – or rather, a shared "not-being-able" to suffer."[41] That not-being-able is directly outlined by Pigeonblog: both birds and humans in the urban environment are vulnerable to the pollutants the project is tracking. Both are mortal biological creatures whose bodies are subject to the impacts of life in the environments they inhabit, to the pathogens their co-inhabitants carry and to the toxins produced from their own acts and technologies.

It is by no means easy to theorize the position of the pigeons as political actors in the project. Can they be attributed willing participation? Can their relationship with their handlers be described as kinship? How could Pigeonblog be reformulated without positioning its human authors at the center of its production? There are a few ways, some suggested by the project itself and some that I will propose, that help us to think through the birds' participation. Da Costa writes:

> With homing pigeons serving as the "reporters" of current air pollution levels, Pigeonblog attempted to create a spectacle provocative enough to spark

> people's imagination and interests in the types of action that could be taken in order to reverse this situation. Activists' pursuits can often have a normalizing effect rather than one that inspires social change. Circulating information on "how bad things are" can easily be lost in our daily information overload. It seems that artists are in the perfect position to invent new ways in which information is conveyed and participation inspired. The pigeons became my communicative objects in this project and "collaborators" in the co-production of knowledge.[42]

In positioning the pigeons as "communicative objects," da Costa marks the inequality of authorship between pigeon and person in the production of art. But the pigeons are also described as "reporters" and as "'collaborators in the co-production of knowledge." By maintaining this kind of ambivalence between the pigeon as object and the pigeon as actor, da Costa highlights the difficulty of inviting nonhuman collaboration in a political and artistic endeavor. She also highlights the ways in which, for this project at least, participants may depend on animal collaboration to gain certain knowledge. The birds who worked on Pigeonblog had access to a view and a kind of knowledge their human counterparts did not. The human participants needed the pigeons to monitor pollution in the areas where regulatory bodies would not. This relationship of dependence, evident in the project, is one avenue down which Pigeonblog can travel toward an ecological modeling of urban relationships. It is not just for the knowledge they helped gather that Pigeonblog's human actors need the birds. Animal life requires other animal life to sustain itself. There may be no way to directly map the complex ways in which pigeons, as a bird species, contribute to human life in urban environments but their population booms and declines have directly corresponded with human population and behavior and the areas in which they congregate are sometimes areas which humans have made uninhabitable for other life. Pigeons are synanthropes whose relationship to human and other life forms in an urban ecology is codependent.[43] All biological life is, in some sense, codependent, both in and out of the city. Political life, too, includes codependency. The way Pigeonblog situates itself in relationship to the birds that participate in the project is, at the very least, an indication of the contemporary requirement to examine ecologically the forces at work in the urban world and to acknowledge the sometimes catastrophic consequences of human participation in the local and global ecology. Pigeons have a stake in what makes up a city, what pollutes a city, what kind of architecture is built in a city, and so do the other life forms, humans included, that inhabit these environments. Ecology, for Pigeonblog, becomes a critical political as well as biological model.

Pigeonblog outlines a way of imagining relationships between human and nonhuman animals and the technologies animals use that invites otherness in to play by positioning the pigeons as collaborators and underscoring the ways in which the birds are constructed as pathological. As the project claims, "species boundaries are being actively reconstructed on the molecular level."[44] This is particularly true in notions of the pathological. It has become increasingly difficult to separate so-called foreign bodies – microbes, bacteria, pathogens, and the like – from our own. The human creature is never only human. The avian body is never only avian.

Donna Haraway's work in *When Species Meet* details the ways in which even biologically, humans aren't humans without a mass of other nonhuman animal, microbial, and informational counterparts.[45] New medical technologies help to ensure this blurring of boundaries. Implanting nonhuman animal organs in the human body or producing human flesh on nonhuman animal bodies, even the use of animal genetic code to learn things about species other than that of the donor – all of these practices make clear that conceptions of life are moving more and more into a world without discrete taxonomies. The human as a biological beast is never isolated from the otherness of the vast and varied microbes, molecules, viruses, even genetic code that penetrate the porous borders of the body even at the cellular level. Nor is it, as Haraway so joyously underlined earlier in her *A Manifesto for Cyborgs*, isolated from any number of technological apparatuses.[46] When the pigeons in the project wear their packs they are hybrid, even cyborg animals. This positioning of cyborg and hybrid relationships between the pigeon and its pack or the human collaborators and their pigeon coproducers is made all the more potent in its political potential when we look at the factors in the environment the project seeks to monitor.

Haraway talks of Pigeonblog directly in her recent *Staying with the Trouble: Making Kin in the Chthuluscene*. She sees in the project an example of "staying with the trouble," that is, refusing the

> destructive [...] position that the game is over, it's too late, there's no sense trying to make anything any better, or at least no sense having any active trust in each other in working and playing for a resurgent world."[47]

Pigeonblog's own language supports her reading here, eschewing the "how bad things are" model. While I very much agree with this position, the aim of this book is in large part to suggest that *playing* with catastrophe, with how bad things will be, and sometimes, yes, how bad things are is part of the work Pigeonblog and the other projects outlined in this book do. I have already discussed some of the darker speculations of the project. They are not, in fact, counter to staying with the trouble. In some cases you can have it both ways. The catastrophic thinking with which these projects play, the fact that they *embody* play, indicates as much.

Pigeonblog is invested in hybrid relationships and while no attempt is made by the project to directly cross-species boundaries, project coordinators insist on the value of rethinking what these boundaries are, all the while reminding users of the fragile position of the biological body in the environment. The toxins the project tracks in the Southern California air also cross porous bodily borders – not just into the human bodies traversing terrain on the ground, but into the plant and nonhuman animal bodies who also make up the particular ecology into which these pollutants are released. Part of what Pigeonblog asks its viewers to do is understand that ecological participation goes well beyond what pollutants are released into the air and water, what species actors claim membership among, or what technologies they use. What the project tells its users and observers is that participation in these ecologies is a political and ethical as well as biological endeavor. Its success is in consistently reminding users that actions in the city are always multidirectional – borders of all kinds are in a constant state of breach. All of this labor complicates and plays with the pathological in its more traditional conception.

There is, though, also a way in which Pigeonblog wants to understand and bring into the field of knowledge the toxicity in what are constructed as pathological spaces. Marginalized neighborhoods around Southern California, particularly those ignored by the logic employed by regulating organizations, the "hot spots" Pigeonblog hopes to examine, are the pathological others to the urban life as it is imagined by developers and urban planners. Such spots are where the pathological body makes itself most present and also where the embodied experience of urban ecologies is most socially charged.

Pigeonblog does not direct its inquiries or commentaries on the material impact of the technologies it uses on worlds and ecologies outside the scope of the project, though it is quick to point out that communication and transportation infrastructures (which it is very clearly interested in using) are partially to blame for the very pollution the project monitors. Surveillance, unsurprisingly, refuses to acknowledge such impacts. I would argue that by at least engaging in a reformation of the pathological as it marks out some of its sources, Pigeonblog is offering its users a gestural articulation of the bigger picture. It asks users and participants to locate themselves within an ecology and to mark their relationships to pigeons and pollutants, political and technological systems, global and local impacts. It is an exercise in the conceptual dissolution of boundaries between the body and its milieu, between the site specific and the global, between technological and biological life.

One final note on the pathological. Eugene Thacker and Alexander Galloway write that "emerging infectious diseases such as SARS not only operate across different networks simultaneously but in so doing also transgress a number of boundaries. The lesson here is that network flexibility and robustness are consonant with the transgression of boundaries."[48] Their assessment was recently proven once more in the Coronavirus pandemic, which also crossed species as well as geographic boundaries. Part of the success of Pigeonblog is in its picking up and overturning of the pathological. Where the dominant discourse on animal–human relationships is fear of cross-species infection, Pigeonblog sees opportunities for coproduction. While certainly disease crosses species boundaries and pandemics travel through transportation infrastructures, so too can we. The network is available for the purposes of resistant action just as it is for the transmission of a pathogen.

Ecology: Life and lively spaces

The final section of this chapter is devoted to the concept of ecology. The term is sufficiently broad as to keep my analysis open to a number of readings, all of which should be kept in play here and in the book as a whole. Looking at Pigeonblog in particular, I want to highlight the way in which the project is concerned at once with "ecology" in the popularly understood sense (the "natural" world, its human and nonhuman life and that life's interaction with and impact on its milieu) and with the media ecologies, the complex of biotechnical relationships, of urban space in which the project works. These are not two separate ecologies, but rather two comingling dynamics – what we call the "natural" is not separable from the technological nor vice versa. As biologist Daniel Botkin writes,

> Nature as we are coming to know it, is a patchwork of complex systems with many things happening at once and with each system undergoing changes at many scales of time and space. Human beings, far from being alien interlopers who disturb the timeless rhythms of nature, are intrinsic elements of the natural order. Chance events seem to play an important role.[49]

The human production or use of technology is not, thus, something to be parsed from the "natural" ecology of a given environment, as Pigeonblog is aware. The project has an interest in the health of the environment and its inhabitants. It wants to gain knowledge of this environment, that can usefully aid in future projects. Pigeonblog uses various technologies within its urban ecologies to produce knowledge about those ecologies. "Grassroots scientific data gathering"[50] can very well result in altering the very ecology it seeks to know, and the impact of the project on the environment it participates in, as discussed, goes well beyond regional borders, the pollutants it monitors, or the pigeons it sends flying through the air.

Pigeonblog is successful not only in situating itself within an urban media ecology, but also in suggesting that public participation and play can offer new means of producing, and valuing, knowledges about such an ecology, and that such knowledges can be counters to biocapitalization. In its mission statement, Pigeonblog directly points to the sociopolitical elements that contribute to a gap in scientific knowledge about pollution levels in marginalized areas. It also hints at a relationship between military technologies and life that long predates the project:

> Pigeonblog was inspired by a famous photograph of a pigeon carrying a camera around its neck taken at the turn of the twentieth century. This technology, developed by the German engineer Julius Neubronner for military applications, allowed photographs to be taken by pigeons while in flight. A small camera was set on a mechanical timer to take pictures periodically as pigeons flew over regions of interest. Currently on display in the Deutsche Museum in Munich, these cameras were functional, but never served their intended purpose of assisted spy technology during wartime. Nevertheless, this early example of using living animals as participants in early surveillance technology systems provoked the following questions: What would the twenty-first century version of this combination look like? What types of civilian and activist applications could it be used for?[51]

Marking this lineage is a way for the project to remind its users that ecologically situated relationships can, as it were, go both ways. Pigeon coproducers can work for vertical hegemony (helping the military to survey bodies and regions for the execution of war, for example) or work against (if still within) it, as they do for

Pigeonblog. Note, too, that the productive inspiration for Pigeonblog is a failure: pigeons never became spies as the engineer who developed the cameras intended. A multiplicity of possibilities are outlined by the project that are not external to the ecological dynamics of the exercise of power, be it by capital and the state, but that can be activated as tactical response or as modulating mechanisms of control.

The concrete functioning of global capital can easily be described as "ecological" as can practices resistant to its regime – dynamic interrelated systems, networks of power and exchange, mobile layers that adapt to change and alter each other. This has increasingly been problematic for biocapital's critics.[52] Pigeonblog goes to great lengths to expose this problem by investing itself, "in pursuit of resistant action" in the very technologies and genealogies that biocapital uses to exercise power.[53] If the military technology that inspired the project was meant to direct a surveilling gaze on bodies, Pigeonblog turns its gaze upward, and outward to that which biocapital does not particularly want us to see. There can be no question that the contemporary technologies of communication and transportation have had toxic effects on local and global ecologies. Pigeonblog points to this directly, noting that, unlike pigeons, "most flying targets are a source of pollution themselves."[54] Once equipped with the pack, with all its manufactured and transported technology, the birds who participated in the project can also become implicated in production of toxins.

It is worth examining the way Pigeonblog figures life in its own project and its regional ecologies. As elaborated earlier, there is a way in which Pigeonblog understands the bodies it engages as both biological and political life. The lively bodies on the ground and in the air have a critical biological mobility in the work – they interact with their environment, impact it and are impacted by it. The way they participate in the knowledge production of the project invites them to make a political articulation. The point is to revitalize debates about the production and regulation of pollution and to highlight the overlap of the political and the biological not just in this field, but in everyday life.

The pollutants and toxins in the air that the project monitors clearly affect the pigeons and other animal inhabitants in the environment. Citing a study in the Spring 2003 *American Journal of Public Health*, the project reminds us that "air pollution levels in the Los Angeles and Riverside counties region are of high enough magnitude to directly affect children's health and development."[55] But far from being removed from the political sphere, these toxins have distinct geo-political consequences – where they are released, how they are regulated and how they are tracked bring them into the proper sphere of politics. In the regime of biocapital, lively bodies and lively spaces are mobilized in any exercise of power. Once more the counterexample of carceral surveillance is illustrative. The state can use technology to monitor not just where the lively bodies they track go, but also what substances they intake. It is a means by which to regulate certain layers of the biotechnical ecology of a given body – the spaces that body moves through, what crosses that body's borders. Both Pigeonblog and carceral surveillance are distinctly aware of the correlation between technology, life, and politics. One is built on extracting capital from life and regulating pathological human and nonhuman animal behavior, the other in outlining living alternatives to this instrumental logic through participation and play.

Pigeonblog looks for biovalue that exceeds the extraction of capital from life in the forms of pleasure and knowledge production its work can help create. Thacker and Galloway write: "There are (at least) two meanings of the phrase 'life-resistance': (1) life is what resists power, and (2) to the extent that it is co-opted by power, 'life itself' must be resisted by living systems."[56] When power, as in the case of BI Incorporated, co-opts life itself it mobilizes life, instrumentally, for profit. It aims to control ecological relationships, even as it utterly effaces the potential damage it may do to those relationships. Pigeonblog outlines the possibilities of a counterforce. It positions itself as a living system, highlighting the varied forces of life-as-living-system and pushing against the reduction of life biocapital makes.

Thacker and Galloway continue to make clear that the system is not only about human life, political, social, or biological:

> Perhaps what is most instructive about this view of biopolitics and resistance is that life-resistance is not exclusive to human agencies and actions, especially when considered from the perspective of networks as living networks. Life-resistance puts forth the difficult, sometimes frustrating proposition that 'life' is not always synonymous with the limited cause-and-effect relations usually attributed to human agencies; in this sense, networks–or living networks–contain an anonymity, a nonhuman component, which consistently questions common notions of action, causality, and control.[57]

In its work, Pigeonblog very much articulates and invites this consistent questioning. Its invitation to the pigeons, to participants who track the project on the blog, and to the technologies it harnesses and even the ecologies in which it tracks pollutants all indicate a system that cannot be comfortably taxonomized, much less controlled. I would argue, too, that ecology is an appropriate term to describe what Pigeonblog wants to point out in both its social and material environment. More than a system, or even a network, ecology demands a level of instability within its workings. All of the relationships bound within them are contingent, precarious even. Ecologies cannot be wholly mapped and they are not predictable: each participant can impact its neighbor, each incoming animal or ideology is changed by the environment it enters. These are the kinds of relationships Pigeonblog and this writing want to map – technology, pathology, and ecology are not separable concepts for the function of biopower, nor for those who seek to harness it toward ends other than the extraction of biovalue.

In the air in L.A.

In the opening lines of Part III of *The Practice of Everyday Life,* de Certeau describes "seeing Manhattan from the 110th floor of the World Trade Center":[58] "To be lifted to the summit of the World Trade Center is to be lifted out of the city's grasp," he writes. "One's body is no longer clasped by the streets that turn and return it according to an anonymous law."[59] De Certeau uses this view from above and the pleasure it affords him to begin to discuss the possibilities of intervention from below, particularly those of the pedestrian who traverses the city against the force of the organized whole. "Ordinary practitioners of the city live "down below," below the thresholds at which visibility begins."[60] The objects of this chapter's investigation support de Certeau's thesis that the city can offer itself in some kind of readable, if still unstable, form to an aerial viewer. For both surveillance and Pigeonblog, this knowledge is more than a fiction produced by the spectacle of a bird's-eye view. Real knowledge and with it, real power, can be gleaned from above. It is no coincidence that mapping applications used for surveillance offer a bird's-eye view of the areas in which clients travel. While the urban pedestrian at street level engages in tactics that work against the city planner's top-down strategies, Pigeonblog suggests that tactical interventions can also come from above. This is particularly relevant to Los Angeles.

As Edward Soja describes so well in *Postmodern Geographies*,[61] the air above the city is a long-contested space. From the early 20th century on, Los Angeles has been the site of major aerospace investment and industrial production. Huge swaths of land near the enormous Los Angeles International Airport are the stomping grounds of corporate giants whose concern has been extracting capital from and ruling the air: again, Parks' vertical hegemony. Though the aerospace industry has suffered a slump that began in the late 90s and continues today, the city remains among its centers. It is also a city, as Pigeonblog points out, in which air quality and pollution (some directly caused by the aerospace industry) are major ecological and political issues. Life above Los Angeles has drastic impacts on life below.

Thinking of the city in the air as well as on the ground takes us closer to a model of urban practice that understands and can account for a multiplicity of actors: pollutants and pigeons, artists and corporations, cell-towers and satellites. The "flexible flows and counter-flows" Deleuze begins to describe have to run vertically just as well as horizontally and, for that matter, in any and all directions. Parks asserts that vertical hegemony can only operate if it returns to life on the ground, It "must be communicated through and materialized as part of culture on earth."[62] If Los Angeles is always imagined only from below, as a sprawl city to be seen through an automobile windshield, a huge number of actors who build its particular urban media ecology are erased, and vertical hegemony is rendered invisible.

Finally, where surveillance perpetually defers catastrophe (the biological catastrophe that could come from the pathological animals it tracks, the social catastrophe of violent crime, the economic catastrophe that comes if we fail to employ their world-ordering technology), Pigeonblog highlights that catastrophe *is*, that for some bodies in our urban ecologies it is a state of unfolding rather than an event. If we are to imagine a more sustainable, equitable future for the cities of the Americas, we, now more than ever, must cling hard to this particular truth. There is little hope of addressing a catastrophe that can never come into discourse as it is, when it is. Pigeonblog lets the catastrophe walk right in. It plays within and around it. Such work is essential, and hopeful.

Notes

1 Image courtesy of Robert Nideffer.
2 Graeme Wood, "Prison Without Walls," *The Atlantic* September (2010). http://www.theatlantic.com/magazine/archive/2010/09/prison-without-walls/308195/.
3 "Interspecies Coproduction in the Pursuit of Resistant Action: A Project by Beatriz Da Costa with Cina Hazegh and Kevin Ponto," *Pigeonblog* (blog), n.d., http://beatrizdacosta.net/Pigeonblog/stills_testflights.php. Accessed June 14, 2013. Image courtesy of Robert Nideffer.
4 "Interspecies Coproduction," http://www.beatrizdacosta.net/Pigeonblog.php.
5 BI Inc. no longer sells cattle tracking devices, but the market continues to exist. Many corporations sell RFID tracking tags to ranchers. Among them are Texas Instruments, which sells its cattle tracking RFID devices for the express purpose of instrumental control of life and extraction of capital from it: "When livestock is consistently well-fed and easy to locate, the value of each animal is maximized both in terms of market timing and the data that accompanies the carcass." "Animal Tracking with RFID" http://www.ti.com/rfid/shtml/apps-animtracking.shtml.
6 Kaushik Sunder Ragan, *Biocapital: The Constitution of Postgenomic Life* (Durham: Duke University Press, 2006).
7 Davis, *Ecology of Fear*, 77.
8 Davis, *Ecology of Fear*, 80.
9 "Animal Tracking with RFID," http://www.ti.com/rfid/shtml/apps-animtracking.shtml.
10 For additional distinctions between these two forms, which I take from the anonymous collective, uncertain commons, see chapter 1.
11 "Global Positioning System. U.S. Government office of National Space-Based Positioning, Navigation and Timing Coordination," http://www.gps.gov/. Accessed June 14, 2013.
12 "Global Positioning System," http://www.gps.gov/
13 "Passive GPS Offender Tracking," www.bi.com/exacutrack. Accessed June 14, 2013.
14 "Interspecies Coproduction," http://www.beatrizdacosta.net/Pigeonblog.php.
15 "City Rankings," 2019 State of the Air Report, https://www.lung.org/our-initiatives/healthy-air/sota/city-rankings/most-polluted-cities.html. Accessed April 29, 2019.
16 Lisa Parks, *Rethinking Media Coverage: Vertical Mediation and the War on Terror* (New York: Routledge, 2018), 3.
17 Parks, *Rethinking Media Coverage*, 4.
18 Laura Kurgan, "You Are Here: Information Drift," *Assemblage* 25 (1994), 18.
19 "Improving the Civilian Global Positioning System (GPS)," http://clinton3.nara.gov/WH/EOP/OSTP/html/0053_4.html. Accessed June 14, 2013.
20 "Improving the Civilian Global Positioning System (GPS)"
21 Robert G. D'Eon et al., "GPS Radiotelemetry Error and Bias in Mountainous Terrain," *Wildlife Society Bulletin*, 30, 2 (Summer, 2002), 430-439.
22 "One Piece Active GPS Offender Tracking," http://www.bi.com/exacutrackone. Accessed June 14, 2013.
23 Beatriz da Costa, Personal interview with the author. Los Angeles, May 11, 2010.
24 Philip E. Agre, "Surveillance and capture: Two Models of Privacy," *The Information Society* 10, 2 (1994), 101-127.
25 Ted Kane and Rick Miller, "Cell Structure: Mobile Phones," in *The Infrastructural City: Networked Ecologies in Los Angeles* (New York: Actar, 2009), 151.
26 "Rule the Air," http://phones.verizonwireless.com/ruletheair/. Accessed June 14, 2013.
27 "T-Mobile," http://www.t-mobile.com/. Accessed June 14, 2013.
28 In this it is very similar to Miami. (See Chapter 3).
29 Norman Klein, *The History of Forgetting: Los Angeles and the Erasure of Memory* (London: Verso, 1998), 49.
30 da Costa, personal interview.
31 Matthew Fuller, *Media Ecologies: Materialist Energies in Art and Technoculture* (Cambridge: The MIT Press, 2005), 166.

32 Eva Horn, *The Future as Catastrophe: Imagining Disaster in the Modern Age*, trans.Valentine Pakis (New York: Columbia University Press, 2018), 43.
33 Nicole Shukin, *Animal Capital: Rendering Life in Biopolitical Times* (Minneapolis: University of Minnesota Press, 2009), 182.
34 "The Janus face of the longing for interspecies intimacy," writes Shukin, "is the horror of breached species barriers, a horror closely bound up with intolerance for the pathological substance of ethnic alterity that Zizek sees lurking in the ideal of multiculturalist tolerance," (Shukin, *Animal Captial,* 204).
35 "Interspecies Coproduction," http://www.beatrizdacosta.net/Pigeonblog.php.
36 "Animal Tracking Technology with RFID," http://www.ti.com/rfid/shtml/apps-anim-tracking. shtml.
37 "Interspecies Coproduction," http://www.beatrizdacosta.net/Pigeonblog.php.
38 Shukin, *Animal Capital,* 189.
39 "Interspecies Coproduction," http://www.beatrizdacosta.net/Pigeonblog.php.
40 da Costa, personal interview.
41 Shukin, *Animal Capital,* 223.
42 "Interspecies Coproduction," http://www.beatrizdacosta.net/Pigeonblog.php.
43 Courtney Humphries' *Superdove: How the Pigeon Took Manhattan* (New York: Harper Collins, 2008) offers a layman's read on the pigeon. She tracks a number of population booms in pigeon and human populations throughout the nineteenth and 20th centuries.
44 "Interspecies Coproduction," http://www.beatrizdacosta.net/Pigeonblog.php.
45 Donna Haraway, *When Species Meet* (Minneapolis: University of Minnesota Press, 2008).
46 Donna Haraway, *Simians, Cyborgs, and Women: The Reinvention of Nature* (New York: Routledge, 1991), 149-182.
47 Donna Haraway, *Staying With the Trouble: Making Kin in the Chthuluscene* (Oxford: Oxford University Press, 2011).
48 Alexander R. Galloway and Eugene Thacker, *The Exploit: A Theory of Networks* (Minneapolis: University of Minnesota Press, 2007), 91.
49 Daniel B. Botkin "A New Balance of Nature," *The Wilson Quarterly* 15.2 (1991), 70.
50 "Interspecies Coproduction," http://www.beatrizdacosta.net/Pigeonblog.php.
51 "Interspecies Coproduction," http://www.beatrizdacosta.net/Pigeonblog.php.
52 Shukin, in her introduction to *Animal Capital* offers and exceptional review of the debate in critical theory.
53 "Interspecies Coproduction," http://www.beatrizdacosta.net/Pigeonblog.php.
54 "Interspecies Coproduction," http://www.beatrizdacosta.net/Pigeonblog.php.
55 "Interspecies Coproduction," http://www.beatrizdacosta.net/Pigeonblog.php.
56 Galloway and Thacker, *The Exploit,* 78.
57 Thacker and Galloway, *The Exploit,* 81.
58 Michel de Certeau, *The Practice of Everyday Life,* trans. Steven Rendall (Berkeley and Los Angeles: University of California Press, 1984), 91.
59 de Certeau, *The Practice of Everyday Life,* 92.
60 de Certeau, *The Practice of Everyday Life,* 93.
61 Edward W. Soja, *Postmodern Geographies: The Reassertion of Space in Critical Social Theory* (New York: Verso, 1989), 204-230.
62 Parks, *Rethinking Media Coverage,* 3.

Interlude 3

Messy/terrain

The city is messy terrain. This is true in multiple registers. Urban centers are the sites of immense waste production, concentrated pollution, unruly aesthetic production, and of social messes. They are the sites of messes to come (social and material), and they are palimpsests of the messes we have already made.

"Mess" and "messiness" have negative connotations (particularly for urban planners, architects, for municipal governments), and some of the urban messes speculated upon in this book have negative, even potentially lethal consequences for city-users: police violence against people of color, large-scale inundations driven by anthropogenic climate change, pollution, disease, etc. But messiness is also part of what makes the city desirable. And messiness may also provide a pathway for a new urban politics.

Throughout the twentieth and twenty-first centuries, theorists of urban life could be divided between those who believed in and advocated for a certain kind of messiness, and those whose aim was a more totalizing tidiness and order. Perhaps the most emblematic figures of this debate in the twentieth-century U.S. are Jane Jacobs and Robert Moses. Jacobs wanted to preserve some of the alternative orders available in the mess of a mid-century New York City. She saw in architecture and urban design, and of course in city planning, a failure of approach, an unwillingness to deal with or understand what she called "organized complexity."[1] Moses, on the other hand, wanted to clean up some of the messes Jacobs found so compelling in the city. His expressway plans were certainly a way to tidy (albeit only for the automobile) Lower Manhattan. Jacobs ultimately won that battle with Moses and the expressway never materialized. Her contributions to urban theory and development practices were enormous. Among them was the ability to see what was, in fact, organized in the mess and to advocate for the essential vibrancy such mess brings for cities and their users.

Today's Moses and Jacobs avatars come in many forms, though the ways in which they understand messiness or don't, aim for ordering the city or jamming its networks, frame it as an emergent system or an unending chaotic process, overlap and intertwine in novel ways.

Messy urban technologies

From a scientific or social scientific perspective, another word for messiness might be complexity. The systems approach to cities, for example, like the work done by Luis Bettencourt, attempts to understand through quantitative analysis a wildly broad range of interacting systems at work within the urban world.[2] Bettencourt's research has poured through enormous data sets about an impressive number of cities to try to map the ways that a host of urban systems interact. He and others like him have found some impressive (if sometimes already intuitively understood) patterns. Given that his work frames the city as essentially a mass of interlocking and interrelated systems, he might initially seem more aligned with tidiness than mess. But viewing the city as a knowable total object is more the popular perception of quantitative research of this kind, and less what such work actually does. Systems approaches seek out what *can* be measured in inherently complex arenas, but they do not claim to measure *all* things in the city. They seek patterns in available data in order to develop a useful model or collection of insights of how various interacting processes within cities operate. Urban planners and developers, among others, have a tendency to oversell the explanatory power of such research, selectively offering statistics to support a given agenda. Bettencourt, for example, has found that GDP and urbanization are closely correlated. While this kind of statistical relationship is frequently cited, these observations are almost never paired with another of the conclusions he and his team reached: that increasing economic productivity, while observably paired with increasing urbanization, also consistently leads to rises in inequality.[3] Systems approaches also have a tendency to leave unexplained and uninterrogated the ways in which cultural phenomena shape what data are gathered and how. Bettencourt and his team have pointed out, for example, that data on the real priorities of the worlds' slum populations are woefully rare.[4]

The "smart city" movement, which privileges computation and big data in urban development and governance, is among those urban approaches that are married to a vision of urban tidiness. Urbanist Adam Greenfield's *Against the Smart City* makes such an argument rigorously. His writing makes visible the legacy of the dueling vectors at work in the city, that is, the strategic drive toward order, mechanization, whole-scale planning and development, and the counterforces of messiness, shock, disorder, and unruliness. Greenfield makes clear that there is legitimate danger in the smart city discourse, in part because of its refusal to contend with the complex messiness of cities as they are lived. He reads material put out by companies like Cisco and Siemens, who are among the largest purveyors of smart city software and, thus are its primary benefactors and most vocal proponents, and finds "the bold claim of perfect knowledge"; this, he argues, is "incompatible with the *messy* reality of all known information-processing systems, the human individuals and institutions that make use of them and, more broadly, with the world as we experience it" (my emphasis).[5]

Crucially, the algorithms on which smart city software relies are not produced in some kind of computational, human-free vacuum, but rather are made within the milieu of, and trained on, culture. We need not look particularly far to see the ways such algorithms could reify any number of extremely damaging ideological discourses, including racism, sexism, and classism. Sofia Noble's *Algorithms of Oppression* meticulously documents exactly this process to expose what she terms "technological redlining."[6] Redlining is, of course, the long-running complex of systems that deny services through various means to people of color, particularly evident in urban space. Here Nobles applies the term to the tightly guarded algorithms behind Google's PageRank search protocols and AdWords tool and shows how they are deployed to perpetuate and potentially foment oppressive ideologies. Redlining is socio-spatial, and the smart city's reliance on the algorithm is, without serious interrogation and a massive overhaul, almost certain to perpetuate, reproduce, and potentially worsen segregation by class and race in the urban environment. This is not the mess we want.

Noble and Greenfield, taken together, make very clear the dangers of a technological approach to the city that does not take into account the ways ordering systems (technological or otherwise) are not, in fact, without their own messy sociocultural baggage. Ash Amin and Nigel Thrift go further, arguing that the contemporary city cannot be fully addressed by approaches that position the city as a system at all. Systems, they write, "so often imply that there is an immanent logic underlying urban life."[7] But as they and an increasing number of urban scholars from across disciplines argue,

> cities are certainly not systems with their own internal coherence. The city's boundaries have become far too permeable and stretched, both geographically and socially, for it to be theorized as a whole. The city has no completeness, no centre, no fixed parts. Instead, it is an amalgam of often disjointed processes and social heterogeneity, a place of near and far connections, a concatenation of rhythms; always edging in new directions. This is the aspect of cities that needs to be captured and explained, without any corresponding desire to reduce the varied phenomena to any essence or systemic integrity.[8]

In other words, the city–both as lived material site and object of study–is a mess.

In the longer history of urban theory, scholars have frequently envisioned the city as embodying in its very means of being the friction between a push toward messiness and strategic attempts at tidiness. For Walter Benjamin, the emergence of the city corresponds to a rise in the shocks of the unruly crowd, on the one hand, and the mechanization characteristic of the rise of capitalism on the other. Michel de Certeau's notion of the strategy (the top–down, view from above afforded to those who design and govern the city) lands on the side of the tidy; his tactics (the work done by the pedestrian and the individual city practitioner) would be in the terrain of mess. For Henri Lefebvre, the dialectical relationship between heterotopy (the ambiguous, the marginal, the different, the simultaneously outside, and interwoven) and isotopy (the rational, mappable, and contained) in the city can be reframed as one between messy and tidy.

Urban theorists contending with what seems a perpetual, even existential tension between messiness and tidiness, have their counterparts and sometimes comrades in activists who use the city to stage, engage, and attempt to alter urban life. Political demonstrations across the Americas and globally have often leveraged a certain kind of mess-making: the black block is perhaps the most visible tactical disordering, but other acts of civil disobedience have similar effects. On November 11, 2016, for example, Black Lives Matter protesters in Miami shut down Interstate 95 and the MacArthur Causeway at 6 pm, at the height of the notoriously bad rush hour traffic.[9] They made a mess, in other words, of many urban commutes. The *cacerolazo* protests utilized in Buenos Aires after the 2001 financial crisis created an incredible din, making an aural mess that echoed for miles from the Plaza de Mayo.[10] In 2017, thousands swarmed Los Angeles's LAX, shutting down traffic flows to and from the airport and confronting riot police, to protest the Trump Administration's proposed travel restrictions on citizens from majority-Muslim countries.[11]

These kinds of tactical responses to the urban underline the power of mess-making and its persistant relevance. They also gesture to the ways mess-making is bound up with urban pleasures. According to participant accounts, *cacerolazos* are joyous experiences.[12] So are lots of ways of moving through and being, or being together in the city. Wandering is one such pleasure, and one with a very long aesthetic history: the roaming flaneur found in the poetry of Baudelaire is certainly, as Benjamin knew, able to leverage pleasure from the urban mess. Rebecca Solnit has written, in her characteristically beautiful way, about the potency of walking in the city. She writes that

> Cities have always offered anonymity, variety, and conjunction, qualities best basked in by walking: one does not have to go into the bakery or the fortune-teller's, only to know that one might. A city always contains more than any inhabitant can know, and a great city always makes the unknown and the possible spurs to the imagination.[13]

The city's charms are consistently wrapped up in its messiness, its unruliness, and its unpredictability and impossible juxtapositions. Thrift has even made the case that some of our messier human dispositions are central to the enjoyment of urban life. As I briefly discussed in Chapter 2, he argues that misanthropy, for example, can be tied not only to fear but to pleasure. And Thrift is deeply committed to the preservation of the mess which, as has long been the case, is often in play with or against the ordering systems of the city.

Dreamworld and catastrophe

Benjamin was among the first to exhaustively address the urban mess. His reading of Baudelaire, his seminal essay on the reproducibility of the work of art, *The Arcades Project*, and his other writings position city life, commodity capitalism, architecture, and art in urban terrain as in many ways made by the push and pull between ordering phenomena and messiness. His interest in the political possibilities of distraction (most famously in the ways in which the masses take in both cinema and architecture) as well as his discussion of shock as an experience specific to daily city life suggest that the messiness of the urban world may well be the key to building an alternative politics. Benjamin recognized, though, that messiness is not in and of itself revolutionary. It too can be leveraged in dangerous, even fascist political efforts. His dialectical images bear this out. As Susan Buck-Morss succinctly writes:

> Heaven and hell; phantasmagoria and shock; dreamworld and catastrophe. These polarities circumscribe the field of Benjamin's images of the modern city and account for their critical, political charge. As dialectical images, they are constructions out of extremes, functioning both as revolutionary inspiration and as political warning.[14]

Reading Benjamin, Buck-Morss worries that urban life has become more and more a cultural and aesthetic milieu with even less radical capacity than the phantasmagorical spectacles of early twentieth-century capitalism. Those spectacles were, at least, the sites of utopic imagination. Today, she writes,

> Utopian fantasy is quarantined, contained within the boundaries of theme parks and tourist preserves, like some ecologically threatened but nonetheless dangerous zoo animal. When it is allowed expression at all, it takes on the look of children's toys—even in the case of sophisticated objects—as if to prove that utopias of social space can no longer be taken seriously; they are commercial ventures, nothing more.[15]

For her, the city as a modern aesthetic project may have been evacuated of its political potential, even in its most celebrated architecture and art. Contemporary media technology, the new digital image, has pushed the city into a space of catastrophe from which there may be little left to salvage. "Benjamin insisted: 'We must wake up from the world of our parents.' But what can be demanded of a new generation, if its parents never dream at all?"[16] The persistent imagination of catastrophe in the contemporary city, however, may be one possible answer to her question. If the utopian imagination fails in the urban milieu, perhaps speculation on catastrophe, in aesthetic terms, in the material city, and in the ways in which we imagine its unfolding in the future, offers an alternative vision and carves an opening for different futures, if not utopias.

Utopias, after all, are usually too neat. This was particularly true of the phantasmagoric spectacles of which Benjamin and Buck-Morss write. But even more radical utopian endeavors, those available to us through speculative fiction, for example, are insufficient. As Jameson points out, they tell us much about our present but offer little in terms of imagining a legitimately, radically different future.

A utopian imagination (however successful) cannot fully maintain such untidy pleasures. Utopian narratives, even when they include frictions and foibles, are circumscribed. But conflict, misanthropy, protest, argument, upheaval, and perpetual becoming: these are essential to the city, to its pleasures, and its possibilities. If we are bound to cities, if they are ciphers for our desires and our fears, they are so precisely because of their messiness. The imagination of disaster within the city, perhaps particularly when it takes the form of a material intervention in it, begins with the mess and participates in particular forms of mess-making. For this reason, it has radical possibilities lacking in imagined utopias, whether capitalist or the work of radical speculative fiction.

Rem Koolhaas has written extensively (though often enigmatically) of the contemporary city. His theoretical work (not, perhaps, the architectural and design work of his OMA office) has also been attracted to and defensive of messiness. And it is with him that I'd like to close this interlude:

> If there is to be a 'new urbanism' it will not be based on the twin fantasies of order and omnipotence; it will be the staging of uncertainty; it will no longer be concerned with the arrangement of more or less permanent objects but with the irrigation of territories with potential; it will no longer aim for stable configurations but for the creation of enabling fields that accommodate processes that refuse to be crystallized into definitive form … it will no longer be obsessed with the city but with the manipulation of infrastructure for endless intensifications and diversifications, shortcuts and redistributions—the reinvention of psychological space.[17]

The project of a new urbanism, for Koolhaas, is a messy one that refuses definitive form. It pushes toward the process and away from stasis. It seems to me that the forms of imagining disaster embodied by the case studies in this book, by BAL, by YOUR, by PigeonBlog, when they are successful, do just this.

Notes

1 Jane Jacobs, *The Death and Life of Great American Cities* (New York: Vintage, 1992).
2 See Bettencourt, Luís M.A. "The Origins of Scaling in Cities," *Science* 340 (June 21, 2013), 1438–1441.
3 Christa Brelsford et al., "Heterogeneity and Scale of Sustainable Development in Cities," *Proceedings of the National Academy of Sciences* 114, no. 34 (August 22, 2017), 8963–8968.
4 Brelsford et al., "Hetrogeneity and Scale," 8963–8968.
5 Adam Greenfield, *Against the Smart City*, 1.3, Do Projects (New York: New York University Press, 2013), n.p.
6 Sofia Umoja Noble, *Algorithms of Oppression* (New York: New York University Press, 2018), 2.
7 Ash Amin and Nigel Thrift, *Cities: Reimagining the Urban* (Cambridge: Polity Press, 2002), 3.
8 Amin and Thrift, *Cities*, 8.
9 Jerry Iannelli, "Hundreds of Peaceful Anti-Trump Protesters Shut Down I-95 and MacArthur Causeway," *Miami New Times*, November 12, 2016, https://www.miaminewtimes.com/news/hundreds-of-peaceful-anti-trump-protesters-shut-down-i-95-and-macarthur-causeway-8919257.
10 "Pese al Estado de Sitio. Cacerolazo, Marchas e Incidentes," *La Nacion*, December 20, 2001,sec.Política,https://www.lanacion.com.ar/politica/cacerolazo-marchas-e-incidentes-nid360761/

11 James Queally, Javier Panzar, and Matt Hamilton, "Protesters Block LAX Traffic, Face off with Police as They Rally against Trump's Travel Ban – Los Angeles Times," *Los Angeles Times*, January 29, 2017, https://www.latimes.com/local/lanow/la-me-lax-protest-20170129-story.html.
12 See, again, Jonathan Stern's description of the casserole student protests in Quebec in 2012: https://soundstudiesblog.com/2012/06/04/casseroles/
13 Rebecca Solnit, *Wanderlust: A History of Walking* (New York: Penguin Books, 2000), 171.
14 Susan Buck-Morss, "The City as Dreamworld and Catastrophe," accessed December 17, 2019, http://susanbuckmorss.info/text/the-city-as-dreamworld-and-catastrophe/#fn:1.
15 Buck-Morss, "The City as Dreamworld," n.p.
16 Buck-Morss, "The City as Dreamworld," n.p.
17 Rem Koolhaas, *SMXL* (New York: The Monacelli Press, 1995), 969.

5 Conclusion (or, how I learned to stop worrying and love the city)

As I write this concluding chapter, two incredible catastrophes have very recently and radically impacted cities across the Americas and the world. One is the global pandemic launched by the vast and rapid spreading of the novel coronavirus. Lockdowns and stay-at-home orders brought urban life to a screeching halt. Miami, where I currently write, shut down hotels, shuttered bars and restaurants, and roped off parks and beaches. Traffic in the city, notorious for clogging highways and roads, vastly diminished. Airlines canceled flights and the city changed its pace and its makeup. It stood, however briefly, mostly still. Just as it began to reopen (despite rising positive case numbers), a second catastrophic event intervened in the quiet: the murder of George Floyd by four members of the Minneapolis police department. This sent thousands into the streets to protest white supremacy and the police brutality that attends and sustains it, in Miami and cities across the United States and the globe. It is the most sustained and populous public uprising since 1968.

Both these crises of public health, these catastrophes, the pandemic and brutal extinguishing of Black life at the hands of police, have very much reached into and engaged the vertical field. Since the pandemic's emergence, there has been an endless productions of maps, almost exclusively from a vertical vantage point, that track the spread of the virus. Its transmission across global regions was facilitated by international flight traffic: it was literally carried from place to place in the skies above us. It leveraged infrastructure toward its own replicating ends. The killing of Mr. Floyd, too, and the uprising it spawned also have stakes in verticality: police surveillance, low-flying military helicopters, even tear gas operate within the vertical field. Even the way in which we depict and try to understand the impact of these intersecting catastrophes on cities has a stake in the vertical: the powerful images, first of emptied urban centers, and then of the enormous mass marches flooding through those same, recently deserted public spaces, were very often shot from above.

What to make of these two catastrophes that so wildly reshaped what it felt like to live and be in the city? It is too early, with both the pandemic and the uprising still unfolding, to tell what cities will look like in a year, or a week. The problem with working on contemporary aesthetic interventions is that, of course, the field in which they appear is changing even as they emerge. Just as the intervention of YOUR appeared and vanished, followed by the building that was its canvas, so too do all interventions endeavoring to shape a city that is always, *already* transforming.

But, as I have argued, the capacity for transformation and alteration embodied by the plasticity of the urban environment is precisely why such interventions merit critical attention in the first place: they play a role in narrating what the city is, and what it has been; they point us toward what the city might be becoming.

It is, however, possible to say even now that the pandemic and the uprising as they currently stand only more tightly link catastrophe and the city and point to the rich, dynamic, and multivectorial ways that catastrophe and responses to it play out in the urban world. The 21st Century city is the catastrophe city, and our notion of it must take into account multiple scales and vantage points, must look, in the face of catastrophe, into the air above and the ground below, toward porous borders and bodies of all sorts. As the projects this work has explored make clear, there is hope in imagining catastrophe in the city, in playing with it, in predicting it, and fighting against it. These works can help us picture different, more sustainable, equitable, and pleasurable cities that might rise from the ashes, or better still, might be built before the next catastrophe hits. They mark the many ways the city is vertically constructed, but also sprawled and spread, reaching not only up but out and inward. Buenos Aires Libre, YOUR, and Pigeonblog are preliminary efforts to mark the city as a place that can and does change, and to envision other cities, other ways of inhabiting them together, other ways of playing with and in them.

These projects offer pragmatic models of urban intervention. They can be copied. They can be taken and remade in other cities. The projects explored here and the ways they can be used to speculate about urban remaking are practical, pleasurable ways to begin answering pressing questions posed by urban landscapes and communities without falling into the trap of either prelapsarian narratives of a lost "nature" or a vision of technological liberation from the boundaries of the material world. Thinking of "nature" as an isolated outside is exceptionally difficult to do when you envision urban testing grounds as rich media ecologies. It is also difficult when we know that nature's forces have changed shape, and put our cities at risk, because of the way we, as humans, have participated in the ecologies we help compose.

The urban interventions in this book do work to remind us that limitless and liberating technological evolution is not the teleological end to historical development. They remind us of the consequences of technologies that work to support vertical hegemony, and of social structures that position the value of some lives and bodies above others. They ask us to think on multiple scales, in multiple registers, about the relationships that cities engender and the larger, global flows of power in which they are enmeshed.

Lauren Berlant's *Cruel Optimism* illustrates the other side of what I've considered as urban contention with and relationship to catastrophe. She argues that "across diverse geopolitical and biopolitical locations, the present moment increasingly imposes itself on consciousness as a moment of extended crisis, with one happening piling on another."[1] Certainly, the grieving family and friends of George Floyd, those of the countless other victims of racialized violence, and those mourning the over one million worldwide fatalities from the pandemic indicate a large-scale sense of crisis, of catastrophe. So too do all the interventions outlined in these pages. But where Berlant's "cruel optimism" points to the ways affective attachments to objects of desire (be they concepts, commodities, substances, or cities) become cruel when our continued attachment to them "actively impedes the aim that brought [us] to [them] initially," the catastrophic imagination displayed by the projects here inverts this kind of affective relationship, so apparently common in the present.[2] These works are, instead of cruelly optimistic, playfully pessimistic. They envision dark futures or outline bleak facts on the ground of the present as a means by which to experience pleasure within the city, to project fantasies of a future city that might be radically different from and differently organized than those we currently inhabit. They call out to an impending doomsday in order to live in and make the city more joyous, livelier, and more vibrant with possibilities right now. Such playful pessimism may be one essential counter to the cruel optimism that the contemporary subject seems so frequently to suffer.

Back to Benjamin: Urban art, technology, and reproducible reimaginings

I think of these projects as urban interventions, of course, but also as forms of urban art. In order to think through the larger implications of their work *as* art, I return to one of the 20th century's greatest thinkers of the city: Walter Benjamin. As a means of going forward by looking back, toward the historical roots of the kinds of art that are emerging now within the urban world, I conclude this book by thinking as Benjamin did, of the relationship, perhaps now more tightly bound than ever, between technology and art. His perennially useful "The Work of Art in the Age of its Technological Reproducibility" provides some essential theoretical grounding.[3] He outlines here, with particular a focus on film, the ways in which social and individual experience, particularly of art, changes with the advent of technologies of reproduction. Benjamin is deeply concerned with, on the one hand, new potential forms of (a)perception ushered in by technologies of reproduction and, on the other, with the possible political futures that might be made – very much for the better or very much for the worse – as a result of these emerging modes of experience. It is in this essay that he centrally leverages the concept of the aura and positions an opening for radical, positive change of the experience of art and objects that, via their reproducibility and reproduction, grow distant from the aura, thus unhinging them from their "parasitic subservience to ritual."[4] In Benjamin's reading, tactile and material forms of perception (distraction is here advocated, as is the capacity of film to unveil the optical unconscious) are unleashed by technologies of reproduction. There are two concepts this essay helps unpack that still underline the stakes of contemporary urban practices that harness technologies, both ancient and emerging. These concepts have been central to the whole of the work of this book: speculation and play.

In the case of speculation, Benjamin locates art in a particular relationship to history and to its possible reformation in the future. "To an ever increasing degree," he writes, "the work reproduced becomes the reproduction of a work designed for reproducibility."[5] Reproducibility built into the reproduced artwork, in his formulation, is the condition of a work of art's entrance into the field of politics. Reproduction is too, by its very nature, a speculative capacity. Reproducibility means reproducibility in time, over time. While "it has always been one of the primary tasks of art to create a demand whose hour of full satisfaction has yet to come," according to Benjamin, technologies of reproducibility bring this temporal gesture made by the artwork into the realm of politics in a novel way by engaging the masses in distracted experience, which is tightly bound to their position in the capitalist economy.[6] The industrialized economy produces a mass of laborers engaged in repetitive work and regimented schedules, and funnels these workers back onto the same city streets after their shifts where the cinema awaits them. In part, the entrance of the work of art into politics is, too, an indication of an increase in the speculative nature of the artwork from the advent of photography onward. In a footnote, Benjamin outlines three "lines of development" at the intersection of which "every highly developed art form stands":

> First, technology is working toward a particular form of art. [...] Second, traditional art forms, at certain stages in their development, strain laboriously for effects which later are effortlessly achieved by new art forms. [...] Third, apparently insignificant social changes often foster a change in reception which benefits only the new art form.[7]

Creative interventions in the urban landscape can be seen as emerging from technological developments and gesturing toward new, as yet unrealized future forms of intervention. The case studies in this book are exemplary in this. Buenos Aires Libre emerged only after the Internet was already a global technology and the work of the network is to envision alternative forms of informational exchange as yet only imaginary in the city. Pigeonblog harnesses a form of surveillance never put into practice historically as well as current surveillance technologies such as GPS to gesture toward new forms of cross-species cohabitation and collaboration. The technology of YOUR's intervention is ancient, of course, but its reproduction and transmission online indicates the local's capacity to sprawl in novel ways beyond the boundaries of any given city. Certainly, contemporary art practices that use technologies, broadly conceived, are speculative in the broadest possible sense of the term. They necessarily indicate a future that they help usher into being, and while they may not in any way foresee the consequences of their participation in that future, it is for it that they urgently require critical attention.

The rejection of the aura and authenticity that emerges in the age of technological reproducibility is also an opening in the work of art toward revolutionary potential, according to Benjamin. There is no reason that aesthetic practices of the sort investigated through the lens of urban media ecologies would have any less of a stake in the game of experiential transformations than film did in Benjamin's era. If anything, both the reproductive capacities of contemporary technology are amplified and the disinvestment in and distance from the aura is made greater, particularly in cities: think of an online interface that is reproducible on any computer screen or smartphone, and of the concentration of those screens throughout an urban landscape.[8] The audience to which such an intervention might be delivered is broad indeed. And too, there is a way in which the very broad dispersal of images and information engendered by the use of contemporary information technologies demands that we rethink verticality in the city: creatively employing the vertical field, as all of these projects in some way do distinctly, is always *also* engaging in other directionalities. Buenos Aires Libre reaches up toward the rooftops and out across the sprawl. YOUR scales to a building's heights and then its image, reproduced, scatters across media networks across the city and into communities outside of it. Pigeonblog employs the orbital (through its use of GPS) to return to the very distinctly local.

Contemporary urban interventions also offer a kind of reproducibility that Benjamin did not forsee. They often encourage users to themselves produce and reproduce the works. Either very clearly, as in Buenos Aires Libre, situated as it is within open source movements and built for and by its users; or loosely, in terms of a project's given frameworks, as when images of the YOUR graffiti was posted on social media, resent, reprinted, even copied on coffee mugs and T-shirts. There is a way in which many urban works *require* the participation of users, observers, passersby to both produce and reproduce the works themselves. Perhaps these are not the masses that composed the audiences of Charlie Chaplin films in the 20th century, but they cannot be imagined as entirely separable from such a crowd if only because, like that crowd, they still have the legitimate potential to come together to shape and reshape the city. Both production and reproduction, thus, are charged in these urban practices by technologies. They also depend on bodies beyond their authors, and, certainly, well beyond any authenticity or aura. I share with Benjamin a deep concern about the direction of such technologies and the works of art they might reproduce might take us. Nonetheless, it is essential to point to the possible escape hatches without losing sight of the forces at the ready to close such routes out and onward.

Play has been another key term in this book and I would like to give a closer reading of it here. Play also concerned Benjamin, who insisted on the value of play in the reproduced and reproducible work of art specifically as a link to political potential. In examining potential revolutionary capacities of experiences enabled by technological innovation, Benjamin explains in two footnotes that the historical move toward technological reproducibility of the artwork opens up new fields of radical play:

> Just as a child who has learned to grasp stretches out its hand for the moon as it would for a ball, so humanity, in its efforts at innervation [collective innervation is the aim of revolution, according to Benjamin] sets its sights as much on currently utopian goals as on goals within reach. For in revolutions, it is not only the second technology [enabling mass reproduction, vs. the first technology of representation and semblance tied to cult and ritual] which asserts its claims vis-a-vis society. Because this technology aims at liberating human beings from drudgery, the individual suddenly sees his scope for play, his field of action [*Spielraum*], immeasurably expanded.[9]

He concludes that "play is the inexhaustible reservoir of all the experimenting procedures" of emerging technologies of reproduction, that, indeed "what is lost in the withering of semblance and the decay of the aura in the work of art is matched by a huge gain the scope for play."[10] While play takes on many forms throughout Benjamin's work – he is interested in children's games and gambling, for example, in *The Arcades Project* – I want to underline how frequently and emphatically in the "The Work of Art" essay he links play to revolutionary potential and how insistent his work is in connecting the expanding possibilities for play to the evolution of technological reproducibility.

Urban play enacted as critical speculation should be read in Benjaminian terms. That is to say, it should be read cautiously, but hopefully. I am invested in thinking about ways such play and its relationship to technologies might prove, if not collectively innervating, at least indicative of the political possibilities afforded by carving out new experiences of lively art in the thoroughly entwined fields of the urban, the technological, and the ecological. It seems to me that because the nature of power in the contemporary moment reaches well into the grounds of life itself, such expansive, hopeful thinking is necessary, odd as it may seem to argue that thinking grounded in the imagination of urban catastrophe is expansive or hopeful. But it is precisely play, the fun that certain forms of speculation, including speculation on disaster, can be in the urban context, that allows what appears to be contradictory to become, rather, comfortably tied together. The urban itself is in fact packed with many echos of such contradiction: dangerous and deeply desirable; dirty and beautiful; chaotic and planned; towering and sprawling.

One final return to "The Work of Art" helps to elaborate the fields in which there is room for play. Benjamin was aware that the work of art in the age of reproducibility (perhaps now, it is in the age of its hyper-reproducibility) has distinct impacts in and on the urban environment and distinct links to urban economies:

> For the majority of city dwellers, throughout the workday in offices and factories, have to relinquish their humanity in the face of an apparatus. In the evening these same masses fill the cinemas, to witness the film actor taking revenge on their behalf not only by asserting *his* humanity (or what appears to them as such) against the apparatus, but by placing that apparatus in the service of his triumph.[11]

The urban figures as key terrain in the 21st-century Americas precisely because of the ways in which social structures, technologies, and bodies (both human and non-) interact within the city to produce fluctuating media ecologies. The early 20th-century cinemagoers in cities across the globe are paralleled by the urban wanderers of the present. Their interaction with the works and practices described can offer insight into the ways in which social phenomena are tied to technologies in historically specific ways. The informational/ecological landscapes the projects studied in this book tackle do engender new ways of knowing and experiencing urban space that are perhaps even more radical than those Benjamin saw in film technologies. In fact, because urban works are distinctly spatial, they might be more akin to architecture as art as Benjamin describes it:

> By contrast [to the concentrated art lover], the distracted masses absorb the work of art into themselves. Their waves lap around it; they encompass it with

> their tide. This is most obvious with regard to buildings. Architecture has always offered the prototype of an artwork that is received in a state of distraction and through the collective.[12]

Whether a user comes across the real-time pigeon tracking in the PigeonBlog project as she surfs the Web, or a driver rests her eyes on the YOUR graffiti as she's stuck on traffic-clogged A1A, participants and viewers of these works may not be active viewers in any traditional sense. A critical look at urban interventions can already find emerging, even in the nonexpert viewer and the passerby, ways of viewing that have critical political potential, that open up possibilities for new relationships between bodies, and that engender new ways of knowing ourselves and experiencing our urban landscapes.[13]

Though these works' invitation to participation is necessarily limited by their geography and the social barriers set up by the framing of each object, I have framed the works here as participatory because they alter, produce, and engage the urban environment as both a material *and* social place. Spaces that are never entirely private are charged by these works. In this respect, urban art and aesthetic practices can interact with a broad range of bodies and things, even those outside the more closely circumscribed audiences or environments of a given work. This is partly why I have argued for sprawl as having critical potential, both as an epistemological strategy to approach the city and as a material phenomenon that has latent possibility for renewed urban politics. Because the projects in the case studies presented here are architectural in some capacity they can, in fact, potentially habituate the city wanderer, even the wanderer who does not remark upon them or participate in them directly, to alternative modes of knowing, making, and narrating the landscape. There may no longer be the class of people that Benjamin called the "masses," but there are certainly crowds and crowded city spaces. There are crowds online as well. And the traces that creative urban projects leave within the cities they work with have much broader impacts than an exploration of any one work can unearth. To frame them as participatory is to suggest that on the one hand, these works demand bodies in space as co-producers, but more broadly, to suggest that they might be casting nets wider than the gallery or the museum, the network or the interface, the performance or the inscription.

Toward urban ecosophy and new cognitive maps

Felix Guattari's 1989 essay, "The Three Ecologies" called for new critical and artistic approaches to contemporary politics. Guattari outlines three ecologies: mental, social, and environmental – which are not discrete entities in and of themselves – to tackle the task at hand: "it is quite wrong to make a distinction between action on the psyche, the socius and the environment."[14] He proposes a new ecological praxis that would seek to "to scout out the potential vectors of subjectification and singularization at each partial existential locus" and to divert the danger of loss of consistency within them.[15] Instead of producing "anxiety, guilt and more generally, psychopathological repetitions," the praxis he calls ecosophy should help usher into being creative repetitions, heterogeneous subjectifications, and lines of flight.[16]

I am not prepared to say that Pigeonblog, YOUR, or Buenos Aires Libre are engaged in this sort of scouting *per se*, but I am sure that they help offer incomplete prototypes of urban thinking more radical and more potentially liberating than those available in what Guattari calls Integrated World Capitalism. By examining both the urban practices outlined here and others like them, there may be a way to begin to pragmatically engender the kinds of subjectification that Guattari advocates.

Urban intervention is also always urban mediation. And its media forms are central to both the possibilities afforded to the urban practitioner within any given intervention and to prohibitions and restraints that prevent that intervention from coming into being. There is not now, nor likely was there ever, an ecology that is not also a media ecology or a city that is not deeply enmeshed in media and mediation.[17] Moreover, of course, the contemporary urban environment is one in which contemporary technologies proliferate wildly. While there are a number of concerns about this proliferation, some of which I have addressed, there remains incredible potential in technologies and infrastructures now present in our cities. I side with Guattari when he writes

> Computerization in particular has unleashed the potential for new forms of 'exchange' of value, new collective negotiations, whose ultimate product will be more individual, more singular, more dissensual forms of social action. Our task - one which encompasses the whole future of research and artistic production - is not only to bring these exchanges into existence; it is to extend notions of collective interest to encompass practices which, in the short term, 'profit' no one, but which are, in the long run, vehicles of processual enrichment.[18]

The works studied here, particularly their use of media technologies, have a deep commitment to processual rather than systematic and systematizing imagination of the urban. They are also concerned with dissensual social action such as Guattari discusses. The horizontal organization of Buenos Aires Libre discussed in Chapter 2 as well as the alternative urban epistemology I locate in its practices is one clear example of how emerging urban media practices and forms of play may well be able to work ecosophically. Furthermore, ecosophy works against vertical hegemony: extending notions of collective interest and thinking in terms of processual enrichment requires the dismantling of certain vertical forms.

Because this book is an invitation and opening to continued thinking on the city in the age of vertical power and catastrophe, the analysis had to rely on partial maps and there are no doubt lines of argument that do not move toward definitive conclusions. Such an exploratory writing practice has its own ecologic – pointing to other fields without fully entering them, crossing disciplinary boundaries in order to blur the lines between them, offering questions that might forever remain unanswered but that, nonetheless, serve to elicit critical response, and a critical making and remaking of the urban landscape and of its situation within global capitalism. If Guattari is correct in his claim that "we need new social and aesthetic practices, new practices of the Self in relation to the other, to the foreign, the strange,"[19] the urban landscape can be one space that is particularly exciting as a potential site for this production. As a means by which to articulate this more directly, let me turn briefly to another work on another sort of ecology: Rayner Banham's 1971 *Los Angeles: The Architecture of Four Ecologies*.

Banham's look at the city was unusually attentive to the kind of built environment that had fallen outside the purview of art and architectural history, and to the infrastructure that makes Los Angeles. (His attention to the hamburger stand, the gas station, the "fantastic" architecture of the city has remained compelling and unique in its critical engagement with the popular and the vernacular as valuable cultural phenomena). His grounding of *The Four Ecologies* in terms of both infrastructural and "natural" environments and their histories is, too, yet another prototype for ecosophical thinking about the city. While Banham's work is concerned with Los Angeles, I would suggest that all cities have (and cannot ever effectively or totally contain) the potential to open toward ecosophy. Guattari suggests as much: "What we must emphasize here is that the work of locating these existential refrains is not the sole province of literature and the arts – we find this ecologic equally at work in everyday life, in social life at every level, and whenever the constitution of an existential Territory is in question."[20]

The city is *the* contemporary site of the everyday – not to the exclusion of other sites but in coproduction with them. William Cronon's *Nature's Metropolis* insisted on this coproduction when it argued that Chicago was as much made by its infrastructural links to its rural surroundings and the agricultural production that sustained it as it was by its own internal urban blossoming.[21] The relationship between Buenos Aires, Miami, and Los Angeles and their cultural and regional contexts suggests as much as well. The increasing fact of urbanity as the locus for social, cultural, economic, and subjective production indicates these landscapes as exactly the sites in which the most productive forms of intervention can and do come. But the urban is also (and of course) increasingly the site of the exercise of power by governmental and economic regimes – the urban is dynamic, mobile. It can move and be moved in multiple political, social, and spatial directions.

In his introduction to *Four Ecologies* Banham writes that "the language of design, architecture, and urbanism in Los Angeles is the language of movement. Mobility outweighs monumentality there to a unique degree."[22] Monumentality relies, of course, heavily on verticality. The Buenos Aires, Miami, and Los Angeles interventions this book explores affirm mobility and refuse monumentality. This move toward the mobile is exercised by the projects both in terms of urban space, technological infrastructure and urban architecture, ecology, and as a conceptual and critical territory with distinct speculative politics. The multimedia relationship between birds, landscape, industrial pollution, and art production fostered by Pigeonblog; the precarious and enormous effort of YOUR's author, and the alternative imaginations and horizontal fluctuations of Buenos Aires Libre offer maps of the urban experience that are dynamic and mobile and understand themselves as such. They seek to highlight mobility itself as productive and distinctly urban.

But thinking about contemporary cities and their relationship to technologies of mobility and speed has too often come at the expense of a direct examination of slow time, concrete and concretely experienced space, and the ecological impact of urban development and its infrastructures and technologies. Mobility and speed, too, require more attention. They are concepts that are lived as well as described. My hope is that this book has done something to both unearth the nature of speed and mobility in these three cities and to problematize constructions of instantaneity and virtuality that derail or efface the very material facts of urban experience, as it is policed, reformed, directed and allowed, or ignored in the cities of today. The maps that the projects studied here make of the city resist the monument, the homogeneous, the stable, the one dimensional, but they do not refuse or ignore the deeply felt material impacts of the forces that work on and through them. Or, if the projects are not yet quite ready or willing to do so, perhaps the media-ecological reading that I have done here will help unearth those often catastrophic material impacts. YOUR does little or nothing to point to the infrastructure on which its circulation beyond the city depended; it has been my job to point out, uncovering that infrastructure in part by tracing YOUR's appearance and disappearance, and by thinking through some of the ways it interacts with and shapes urban space. This book overall has sought not only to expose some of the material presences of these projects in the city and the ways in which those presences are in the process of shaping knowledge and practice, space, and experience in and of the city, but also to position my own participation and production as an essential component of the layered urban media ecologies explored.

This book should be, then, a dynamic and multidirectional cognitive map of urban experience. Each of the urban interventions I discuss shapes for itself, its users, members, and participants, and passersby a picture of the cities in which it works. These are both literal (as in Buenos Aires Libre's real-time map of its networks nodes and their connections) and figural (as in the organization's understanding of line-of-site connections between nodes and its activation of an alternative formulation of sprawl). My writing imposes a critical reflection onto these maps, one that seeks to reveal both the present and practical interventions the projects make and to expose some of their blind spots.

In Jameson's formulation of cognitive mapping, he draws on Kevin Lynch's *The Image of the City*, specifically what Jameson calls "the dialectic between the here and now of immediate perception and the imaginative or imaginary sense of the city as an absent totality."[23] Jameson is less interested in the experience of the city and its imagined totality as it is felt, and more in "the local positioning of the individual subject and the totality of class structures in which he or she is situated."[24] For him, the subject, in her local, site-specific position perceives and imagines her immediate environment and this perception is dialectically related to a global totality: capital. Jameson sees capital as global, enormous in its effects, and itself unrepresentable. The subject can picture at best a rough sketch of the absent whole, a cognitive map that points to capital but cannot fully grasp it. This book might be conceived also as such a map, but it is distinctly more interested in pointing off the page and opening up playful ways of inhabiting the city that might in fact be outside the reach of the ordering forces of capital and vertical hegemony. It may be that they are gestural exits rather than the full fruition of an externality to capital, but even envisioning an outside – even an outside brought on by capital's own catastrophic productions – can be a radical intervention in the goings on of the inside.

Where do we go from here?

Verticality, catastrophe, and the city are engaged in dynamic and always ongoing relationships. And vertical power is currently very much entrenched in urban life. Its consequences have been catastrophic: incredible rises in inequality across the Americas and beyond, renewed and virulent racist discourse and continuing racial violence, dismantled ecological regulation even as the consequences of anthropogenic climate change become increasingly devastating. All this suggests that globally, we need new ways to counter these forces, to reimagine and rebuild the city.

The projects discussed in this book are attempts to reshape the city in more equitable and sustainable ways. They open up possibilities for imagining different futures, futures that don't all end in catastrophe. They are all invitations to intervene, to rethink, to reform, and to refigure our urban worlds, invitations we must accept.

We're in luck, despite everything, because there is a great deal of fun to be had in the dismantling of the powers and discourses that oppress us. Even in our darkest prognostications about the catastrophes to come and even in the midst of those already here, we're lucky because there is room for play. Another city awaits us. It belongs to us all.

Notes

1 Lauren Gail Berlant. *Cruel Optimism* (Durham: Duke University Press, 2011), 7.
2 Berlant, *Cruel Optimism,* 1.
3 Walter Benjamin, *The Work of Art in the Age of Its Technological Reproducibility and Other Writings on Media* (Cambridge: Belknap Press, 2008).
4 Benjamin, *The Work of Art*, 24.
5 Benjamin, *The Work of Art*, 24.
6 Benjamin, *The Work of Art*, 52.
7 Benjamin, *The Work of Art*, 52.
8 Scott McQuire's *GeoMedia* argues that the proliferation of media has, in fact, shaped the whole of public space. Cities are "increasingly media-dense spaces" currently being "extensively recalibrated" by the proliferation of both digital devices and data. Scott McQuire, *Geomedia: Networked Cities and the Future of Public Space* (London and New York: Polity, 2016), 19-20.
9 As the translator notes: "*Spielraum*, in this note [...] literally means "playspace," "space for play."" In this writing *space* remains central. Benjamin, *The Work of Art*, 45.
10 Benjamin *The Work of Art*, 48.
11 Benjamin, *The Work of Art*, 31.
12 Benjamin, *The Work of Art*, 40.
13 I want to note, too, that distraction as Benjamin outlines it is also and absolutely counter to rigid verticality. It is one means of revaluing the political stakes of affective registers and communal habituation that remain (though Berlant's work and others have sought to change this) ignored or dismissed in a hierarchical construction of political engagement and subject formation. Benjamin's body of work, in fact, is quite distinctly positioned against a purely vertical construction of power and politics. His utilization of the constellation, his valuation of the experiential, and his refusal of a narrative history that reads as progress: all take aim at vertical topologies as insufficient at best, and oppressive and myopic at worst.
14 Felix Guattari, *Three Ecologies*, trans. Ian Pindar and Paul Sutton (New York: Continuum International, 2008), 28.

15 Guattari, *Three Ecologies*, 30.
16 This passage appears in Felix Guattari, "The Three Ecologies," trans. Chris Turner, *New Formations* 8 (Summer 1989) not in the Continuum edition from which I have drawn other quotations.
17 For an exhaustive and compelling read on this, see Shannon Mattern, *Code and Clay, Data and Dirt: 5000 Years of Urban Media* (Minneapolis: University of Minnesota Press, 2017).
18 Guattari, *Three Ecologies*, 45.
19 Guattari, *Three Ecologies*, 31.
20 Guattari, *Three Ecologies*, 31.
21 William Cronon, *Nature's Metropolis: Chicago and the Great West* (New York: W.W. Norton & Company Ltd., 1991).
22 Rayner Banham, *Los Angeles: The Architecture of Four Ecologies* (Los Angeles: University of California Press, 2001), 5.
23 Fredric Jameson, "Cognitive Mapping," in *Marxism and the Interpretation of Culture*, ed. Cary Nelson and Lawrence Grossberg (Urbana and Chicago: University of Illinois Press, 1988), 353.
24 Jameson, "Cognitive Mapping," 353.

Index

Page numbers in *italics* refer to content in *figures*.

For Product Safety Concerns and Information please contact our EU
representative GPSR@taylorandfrancis.com
Taylor & Francis Verlag GmbH, Kaufingerstraße 24, 80331 München, Germany

www.ingramcontent.com/pod-product-compliance
Ingram Content Group UK Ltd.
Pitfield, Milton Keynes, MK11 3LW, UK
UKHW022018180926
475701UK00005B/916

* 9 7 8 0 3 6 7 5 1 9 3 6 0 *